Genomics in the AWS® Cloud

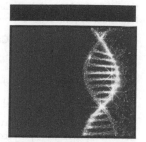

Genomics in the AWS® Cloud

Analyzing Genetic Code Using Amazon Web Services

Catherine Vacher
David Wall

For Roland and Floriane

Acknowledgments

This book represents a lot of work over the course of several years, and we are grateful to the people who helped us complete it.

First among these is Emma Rath, a skilled scientist and good friend who reviewed and contributed to several chapters.

We acknowledge the skill and monumental patience of the team at Wiley, particularly John Sleeva and Kenyon Brown. We appreciate all that they did to bring this book into being.

We are grateful to Carole Jelen, our literary agent, who helped mold this book and also showed extraordinary patience.

We also want to extend special thanks to our family, including Philippe Vacher and Lee and Anne Wall. They were inspirational and always have been.

Thank you, all.

—*Catherine Vacher and David Wall*

About the Authors

Catherine Vacher, PhD, is a research fellow at the Brain and Mind Centre at the University of Sydney. There, she models complex systems related to healthcare and advises policymakers on the most effective use of limited public health resources.

She previously worked as a scientist at the Garvan Institute of Medical Research. There, she worked in bioinformatics, whole-genome sequencing, transcriptomics, molecular modeling, and metagenomics, mostly with respect to cancers.

She also enjoyed a long career at IBM, in Europe and in the Asia-Pacific region. She enjoys theater and opera as well as travel.

David Wall is a consultant specializing in the AWS cloud. His projects have included complex database and machine learning solutions as well as telecommunications systems.

He is an avid cyclist and works as a volunteer firefighter.

Contents at a Glance

Contents

Introduction

Welcome to *Genomics in the AWS Cloud*!

From its title, you can conclude that this book is about two things: genomics (the science of sequencing and interpreting genetic data) and Amazon Web Services (one of the three big hosted computing platforms). *Genomics in the AWS Cloud*, therefore, is meant to appeal either to people from a biology background who want to learn how to do genomics work with AWS or to people with a computer background who want to find out how to apply their skills to genomics.

Both of these areas, genomics and cloud computing, are evolving constantly, and practically no one can claim to be completely *au fait* with either. This book, therefore, aims at not one but two separate moving targets. Our goal as authors is not to teach you everything there is to know about AWS and genomics—or even about the intersection of the two fields—but rather to show you the following:

- Enough of the general concepts of cloud computing and genomics that you understand the problems to be solved and the technologies available to work on those problems

- Enough specifics to enable you to work through actual genomics tasks and see results

Who Should Read This Book

This book is intended for people who aren't content to use commercial genome sequencing services and want to do their own analysis. We walk you through the process of getting raw data from a blood sample via a lab and then using the AWS services to analyze it—learning which genes are present in the sample and what they might say about you and your health. This will enable you to investigate aspects of your genome that commercial services don't explore because they are not allowed to give medical advice.

As well, this book is suited to people who want to learn about the AWS cloud and want to structure their study around a useful field—genomics.

Genomics

At the core of genomics is genome sequencing, which is the process of taking some biological material, such as blood or tissue, and converting it to pure information. This is a complicated process that combines the traditional work of a biologist (which is to say, manipulating actual cells in a "wet lab" environment) with information technology. Cells go into the process; a computer-readable data file comes out.

Genome sequencing took a long time to figure out. Crude, expensive methods were first employed in the 1970s and 1980s. More automated methods became available in the late 1990s, and these enabled the sequencing of relatively simple organisms: yeasts, bacteria, and a nearly microscopic nematode worm (*Caenorhabditis elegans*, long popular in biology labs as an experimental subject). Uncomplicated plants (notably *Arabidopsis thaliana*—a European weed with a particularly small genome) and modest insects (*Drosophila melanogaster*—the fruit fly), both longtime standard experimental subjects, soon followed around the turn of the century.

Two biologists described the first draft human genome sequence in an article in the journal *Science* in 2001. Scientists have worked to refine the human genome sequence since then and also have worked to sequence the genomes of thousands of other organisms.

Key to their work has been a continuous drop in the cost of full-genome sequencing. The first human genome sequence in 2001 cost roughly 2.7 billion U.S. dollars to produce—it required funding of the sort only national governments could provide. Within less than a decade, by 2005, the cost had fallen by four orders of magnitude to something like $1 million—still quite a lot. At this writing, in 2022, it is possible to have a human genome sequenced for less than the cost of a high-end smartphone, and plenty of companies are attracting funding for their plans to bring the cost to less than $100. By the time the asteroid 99942 Apophis makes its closest approach to Earth in 2029, sequencing a full genome will almost certainly cost about the same as the simplest routine medical blood test costs now. The cost of knowing everything about your genetic makeup will be trivial (assuming Apophis doesn't render this unimportant, which the latest reports assure us it won't).

The dramatic fall of the price of genome sequencing, from billions of dollars at the turn of the twenty-first century to a few hundred dollars today, makes it possible for almost all of us to explore our genetic makeup. While we all, as humans, share practically all of our genetic code (upwards of 99.9 percent), the differences make all the difference.

The tiny fraction of our individual genomes that differ from other humans is what accounts for whether we are male or female, all of our physical characteristics, many of our personality traits, and our propensity to health or various kinds of disease.

The availability of low-cost genome sequencing has revolutionized medical and pharmaceutical research and is started to change the practice of medicine. It also enables us to start to understand the building blocks of life and how much, or rather how little, we differ from other life forms.

Genomics in the AWS Cloud is about discovering and studying those differences and learning from them. But there is another part to the equation, which is to say the other set of tools and techniques identified in the title.

Cloud Computing and AWS

Almost in parallel with the advances in genomics that took place between 1995 and the present day, so-called cloud computing evolved enormously during the same time period and today represents a standard way of designing, deploying, and operating information processing systems.

Now, the idea of computing resources that are not local to the people who need them is not new at all. The earliest commercial and scientific computers were, of course, mainframes that were shared across many users—and more than a few of these remain in place today. Servers in organizational or co-location data centers, providing storage and computing resources to privileged users and the general public, have long been part of information technology. In the case of mainframes and client-server systems, users access remote computer systems (often not knowing or caring where they actually are). Functionally, that's *cloud computing*, and it's not a new thing.

What is new is the ease with which modern cloud computing platforms allow rapid construction and cost-effective use of complex and powerful systems. You can quickly set up elaborate workflows, test them with minimal computing power, and then scale them up enormously when it's time for a production run. More or less, you pay only for the computing power you use, and there are ways to schedule the use of processor cycles for times of low demand, when computing is cheaper. With the exception of storage and data transfer—meaning machine images that can be turned into working compute resources, as well as input and output data—systems configured in the cloud can cost practically nothing when they are not doing useful work. Such efficient use of expensive resources isn't possible with on-premises or traditionally hosted solutions.

There are three main players in the cloud-computing industry.

- Amazon Web Services (AWS)
- Google Cloud Platform (GCP)
- Microsoft Azure

Each of them has its points of strength and weakness, and those relative merits are beyond the scope of this publication. Most organizations mix and match parts of each anyway to take advantage of relative technical superiorities and to maintain leverage with vendors. Suffice it to say that we chose to do our genomics work on AWS.

Considering that genome analysis is essentially a workflow to be carried out on storage and computing resources, AWS is well suited to the job. Here are some of the tools we will use:

- Elastic Compute Cloud (EC2) for building and running the Linux servers that actually run the software required for genome analysis
- Elastic Block Storage (EBS) for maintaining updated disk images, ready to attach to a working machine when needed
- Simple Storage Service (S3) for storing input and output files when ready access to them is needed
- Simple Workflow Service (SWF) for automating processes
- Glacier when files need to be archived at low cost
- Identity and Access Management (IAM) for maintaining security and appropriate user privileges

What You'll Learn from This Book

This book is intended to educate its readers in two areas: the science of genomics and the technology of Amazon Web Services. The idea is that you use the latter as a tool to explore the former.

Since it's unlikely that many readers are familiar with both genomics and AWS, this book is meant to teach you either subject—or both if you are familiar with neither—and how they work together.

How This Book Is Organized

Here is a quick introduction to each chapter in this book. You can skip directly to the parts that interest you most, or you can read from beginning to end to get a complete picture.

Chapter 1: Why Do Genome Analysis Yourself When Commercial Offerings Exist?   This chapter explains what turnkey commercial services (such as 23&Me) exist and what they are good for. It then explains what they do not do and why you might want to do your own genomics work.

Chapter 2: A Crash Course in Molecular Biology   This chapter brings you up to speed on the state of biological science as it pertains to genomics. Use this to refresh your knowledge from school or to gain a new understanding.

Chapter 3: Obtaining Your Genome   This chapter explains how to get a sample of your blood suitable for sequencing and how to send it off to a lab for conversion into raw genome data.

Chapter 4: The Bioinformatics Workflow   This chapter approaches the sequencing process from a biological point of view, explaining how one step of data processing feeds into the next to ultimately produce the results you want.

Chapter 5: AWS Services for Genome Analysis   This chapter represents a deep dive into the AWS services you can use for genomics work. If you know biology but aren't familiar with AWS, you might want to begin here.

Chapter 6: Building Your Environment in the AWS Cloud   This chapter goes into more detail about how to set up AWS services for genomics work.

Chapter 7: Linux and AWS Command-Line Basics for Genomics   All major bioinformatics tools run under Linux, so you'll need to understand that operating environment and how to get things working together within it.

Chapter 8: Processing the Sequencing Data   This chapter explains how to get from raw sequencing data to useful information about a genome.

Chapter 9: Visualizing the Genome   This chapter introduces you to tools you can use to depict genomic information graphically, enabling you to understand it better.

Chapter 10: Containerizing Your Workflow on the Desktop   This chapter explains how to use Docker containers, both locally and in AWS, to process data efficiently and scalably.

Chapter 11: Variants and Applications   This chapter explores certain aspects of genome analysis, allowing you to dig deeper into the information you have derived from your sample.

Chapter 12: Cancer Genomics   This chapter discusses the analysis of somatic mutations, specifically those found in cancerous tumors, although the workflow could also be applied to any tissue in the body.

How to Use This Book

You can approach this book by reading straight through, from beginning to end. That is probably the best way to approach it if you have knowledge of neither genomics nor AWS. Alternatively, if you feel you have a good handle on one or the other, you can focus on those chapters that cover your weak area and then study the sections on how AWS can be used to create and automate a genomics workflow.

Our Story

This book is not about us, but it's probably fair to explain to you, our reader, who we are and where we are coming from as we write this book.

We are a married couple and the parents of two children. Catherine is a working scientist and bioinformatician, attached to a medical research group in Sydney. David is an information technology and communications consultant who designs and builds AWS solutions, among other things.

A key fact is that one of our two children isn't alive anymore. Our daughter, Floriane, died of a sudden cardiac arrest in April 2015. She was nine years old at the time.

Floriane's death and its consequences are not the subject of this book. However, the way she lived and died inspired us to take our knowledge of genomics and cloud computing and look inward at our family. We wanted to know what happened to Floriane and what bearing that had on the rest of us. One way we honor her memory is by attempting to understand what happened to her. As well, we want to keep the rest of us—particularly her younger brother, Roland—safe.

Floriane was a fit and apparently healthy girl of nine. She had no ongoing medical concerns and was by all appearances developing normally. She was active and happy.

One evening, after dinner, she approached David on the sofa and asked him to chase her around the house—one of her favorite activities that they had indulged in many times before. David chased Floriane around, theatrically waving his arms. Floriane giggled excitedly and ran a few meters. David caught her and flipped her upside-down, which excited her further because she knew it led to tickling. At that point, she went completely limp.

David was confused. He thought he'd accidentally hit Floriane's head on something. Catherine, nearby, thought it was a seizure, even though Floriane had never had one previously. It quickly became clear that Floriane was in cardiac arrest.

Not having a defibrillator on hand, we called for an ambulance and did our best in the seconds before she died with manual cardiopulmonary resuscitation

(CPR) that proved ineffective. Paramedics showed up with a defibrillator and adrenaline, and used both, but it was all over by then. Floriane died at home, a few minutes after the onset of her cardiac arrest.

(Again, it's not the subject of this book, but please take this opportunity to contemplate the fleetingness of life and to appreciate the people you love. We'll wait.)

And so we were left with questions: What just happened? Why did it happen? Is it going to happen to any of the rest of us? There were many more questions, far more existential in nature, and those questions remain, but those questions are subjects for other venues. Looking at the situation in a limited way, our daughter had just died, completely unexpectedly, and we wanted to know why.

Speculation on the cause of Floriane's death began almost immediately. She hadn't been visibly sick at all. She hadn't exhibited the symptoms—fever and so on—of an infection, such as meningitis. She was far too mature—nine years old, tall, and apparently fit—to have succumbed to the strange and almost random things that claim the lives of babies and are grouped under the catchall descriptor Sudden Infant Death Syndrome (SIDS). She didn't have any known allergies. She wasn't taking any medicines. She hadn't suffered an injury, hadn't eaten anything unusual, hadn't traveled anywhere dangerous—she hadn't done anything outside the realm of what is normal for a fifth-grade girl living in Australian suburbia. And yet she was no longer alive.

The emergency-room doctor offered a broad guess that turned out to be right: cardiomyopathy, or a disease of the heart muscle. Further investigation by the medical examiner showed some characteristics of hypertrophic cardiomyopathy (HCM), which is one of several kinds of cardiomyopathy, but also revealed some characteristics of arrhythmogenic right ventricular cardiomyopathy (ARVC), another disease of the heart. Genetic testing showed two interesting genes: MYH7 (associated with HCM) and RYR2 (associated with ARVC).

The medical examiner officially attributed Floriane's death to HCM but noted that her case had "unusual features"—a reference to the traces of ARVC that had been found. The stated cause of death could just as well have been ARVC, we were told.

We began to research and to think. We discovered that RYR2 is associated with disturbances to the heart rhythm due to adrenaline. Floriane had gone into cardiac arrest when she was wound up with excitement, playing with David. She had received an injection of adrenaline from the paramedics. Could the undeniable structural defects caused by HCM, which usually do prove fatal, but not until age 30 or so, have been amplified by some form of ARVC?

The question gained urgency when we learned that while Floriane's MYH7 was a *de novo* mutation, she hadn't inherited it from us, and her brother didn't have it.

So began our research into our family's genomes.

Getting Under Way

Our goal in *Genomics in the AWS Cloud* is to make the study of your genome, or whatever genome you choose (some are available on the Internet for practice; see Chapter 4), as simple and straightforward as possible. Chapter 2 describes things you should consider before starting. As genome analysis requires computer resources that exceed the average desktop, we set up the analysis environment on the AWS Cloud—that's the subject of Chapters 5 through 7. This book does not presuppose any particular knowledge in molecular biology (which we introduce in Chapter 3) or computers (we talk about Linux in Chapter 8). We paid particular care in explaining how to make sense of our genetic variants in Chapters 11 and 12.

Now, let's start the journey.

How to Contact Wiley and the Authors

Sybex strives to keep you supplied with the latest tools and information you need for your work. If you believe you have found a mistake in this book, please bring it to our attention. At John Wiley & Sons, we understand how important it is to provide our customers with accurate content, but even with our best efforts an error may occur.

In order to submit your possible errata, please email it to our Customer Service Team at `wileysupport@wiley.com` with the subject line "Possible Book Errata Submission."

You can email David Wall with your comments or questions at `david@davidwall.com`.

1

Why Do Genome Analysis Yourself When Commercial Offerings Exist?

As you begin to explore sequencing a genome and analyzing the results, you will no doubt become aware of a number of commercial operations that offer to do the job for you, neat and tidy, in exchange for a modest amount of money. Why would you want to go through the time and hassle involved in doing the work yourself when such convenient offerings are so easy to use? Why not engage a service to do your genetic analysis and enjoy the benefits of something that "just works"?

The answer has, essentially, two parts.

The first is that the commercial services may not provide all the information you want, not least because they are hamstrung by regulations that govern the provision of medical advice. They tend to provide "novelty" information—Larry Lightbulb kinds of things about hair color and finger length, as well as about racial, national, and tribal heritage. They are great for educating children about the sorts of information that DNA can carry and for talking about heritability of characteristics good, bad, and neutral. When you do the sequencing and analysis yourself, you can extract whatever information you want.

The second is that you are the kind of person who likes to do things yourself, either just for the satisfaction of it or because you want to understand how everything works and fits together. Working on your personal genome in Amazon Web Services (AWS) is an excellent way to learn about those services, and that knowledge can then be put to use for fun and profit.

Commercial Sequencing Services

As you survey the Web, you will find that there are several popular consumer-grade genome sequencing services, including 23andMe, Ancestry, and MyHeritage. Others include the following:

- **African Ancestry**: This service is marketed to Africans and people of African heritage. Some users have regretted that the service does not show DNA broken down by origin in the various regions of the African continent.

- **Athletigen**: This service focuses on markers related to physical fitness, such as those affecting endurance and speed of recovery after exertion.

- **DNAFit**: This service provides diet plans designed around certain nutrition-related markers.

- **Fitness Genes**: This service offers training regimes that fit genomic markers related to physical exertion.

- **GEDmatch**: This service focuses on genealogy and links with others who have submitted their sequences.

- **Genome Link**: This service provides information on a series of characteristics, such as physical endurance and skin color.

- **Genopalate**: Focused on nutrition, this service aims to help its customers optimize their diets.

- **Living DNA**: This is an ancestry-focused service with a user community primarily from the United Kingdom and Ireland.

- **MyHeritageDNA**: Ancestry focused, this service connects its users with possible relatives and suggests possible genetic risks to health.

- **Nebula Genomics**: Offering a monthly subscription that entitles its users to monthly updates as new information becomes available, this service includes data on the oral microbiome (i.e., the bacteria found in your saliva).

- **Promethease**: This is a modestly priced service that detects a number of single nucleotide polymorphisms (SNPs, or "snips").

- **Sano Genetics**: This free service concentrates on SNPs related to autism and mathematical reasoning.

- **SelfDecode**: This is a general-purpose detector of several thousand genetic markers.

- **Vitagene**: Focused on fitness and athletic performance, this service includes ancestry information as well.

- **Xcode Life**: This service offers several low-cost specialty tests including one on skin care and another on metabolic diseases.

Typical Results

Of the aforementioned services, perhaps the best-known of the direct-to-consumer genetic testing services is 23andMe, a California company that pioneered the industry in 2007. When someone places an order with 23andMe, the company sends out a kit containing materials needed for the collection of saliva, which is then sent back for analysis. (The idea is that everyone's saliva contains cells that have been shed from the interior of the mouth.) The company presents its report to the customer via its website.

The company ran afoul of the U.S. Food and Drug Administration (FDA) in 2013, when the regulator objected to 23andMe (and other genetics services providers) advertising that its service provided its customers with information on their susceptibility to various genetically linked conditions, such as male-pattern baldness and certain kinds of cancer. This, the FDA said, constituted medical advice of the sort that should be formulated and delivered by a qualified doctor. The 23andMe tests were medical devices and should be regulated as such.

After going quiet for several years, 23andMe applied to the FDA for permission to include information in its reports about a number of mutations and alleles that are well-understood to be associated with pathogenic conditions, including Alzheimer's disease, Parkinson's disease, celiac disease, and a number of BRCA1 and BRCA2 mutations associated with breast cancer. The company argued—and the FDA ultimately agreed—that the test methods used by 23andMe were sufficiently reliable and understood as to not require the involvement of a medical professional. As well, the entities agreed that the relationships between the tested sequences and the various diseases were adequately proven, and that if an individual was found to have a sequence known to be pathogenic, there was no need to hide the truth behind the medical establishment.

With the shift toward presentation of information about ancestry rather than medical conditions, direct-to-consumer online genetic analysis services have, perhaps predictably, begun to appeal to those whose ancestry is more than passing interest.

So, what's in a set of 23andMe reports? If you undergo the saliva test and log into the 23andMe website today, you will get a lot of novelty information about probable hair color, the shapes of certain body parts, and the aspects of aging.

Sometimes, 23andMe gets things right. For example, 23andMe predicted the following for me:

- A 67 percent probability of little to no back hair. (Correct!)

- A 32 percent probability of a bald spot on the top rear of the head. (Correct, pretty much. It's kind of thinning there, but certainly not bald. Certainly not.)

- A 74 percent chance that the earlobes are separate from the sides of the face. (Hear, hear!)

- A 71 percent chance that the ring fingers are longer than the index fingers. (Yep.)
- A 1 percent chance of red hair. (It's brown.)
- A 62 percent chance that dandruff is sometimes a problem. (It is.)
- A 25 percent chance of being afraid of heights. (I am a qualified pilot.)

Sometimes, it gets things wrong. In my case, it forecast the following:

- A 51 percent chance of blue eye color. (They're green.)
- A 66 percent chance of wavy hair. (It's straight.)
- A 33 percent chance of a widow's peak across the front of the scalp. (There is definitely a prominent one.)
- A 4 percent chance of bunions in the feet. (A substantial one on the right foot has been causing trouble since high school.)

From there, the 23andMe report can get a little comical. For example, my report states that I am likely to wake up at precisely 7:34 a.m. Apparently, the company uses statistical analysis of surveys conducted on people with certain similar genetic markers (those associated with early or late rising) to arrive at the precise time. The suggestion is that genetics determine your wake-up time to the minute, which is just not correct. This is a manufactured novelty "fact," and it's not correct: I almost always gets out of bed before 6 a.m. Hilariously, my wife Catherine's 23andMe report has her waking up significantly earlier than I do, which has happened exactly zero times.

The company provides an ancestry report, which attempts to describe which part of the world your forebears came from.

Some background here: my wife and I live in Australia, to which we both immigrated (Catherine from France and me from the United States) around the turn of the 21st century. We are both people of the New World, largely characterized by its relatively recent immigrants, now.

In my case, the ancestry timeline (shown in Figure 1.1) appears to agree with much of what I know about my family history. It shows ancestors from Scandinavia as recently as 1940. Family lore states that my maternal grandmother was born in Sweden and was brought to Chicago as a baby in the 1920s, so that makes sense. Similarly, my family records have my maternal great-great-grandfather emigrating from Hesse, in the western part of what had recently become a unified German Empire, in 1879. That fits with 23andMe's report of French and German ancestry in the late 19th century.

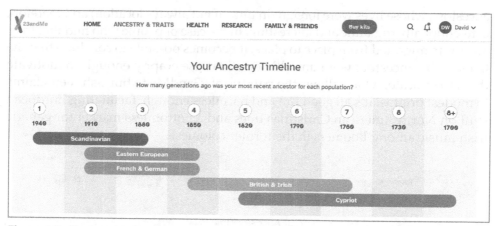

Figure 1.1: My ancestry timeline

My paternal side is less well documented, probably because those ancestors have been in the United States for several generations. The surname Wall fits with heritage from the British Isles, though, and there was always vague talk about my paternal grandmother having ancestors from the eternally contested regions of Central and Eastern Europe.

The report includes Cypriot heritage in the 18th century. I have no idea what that is about, having never heard anything about ancestors on an island of the Ottoman Empire in the Eastern Mediterranean. The 1700s are the distant past for us, as far removed in memory as the original ancestors that first distinguished themselves from the other apes.

Conclusion: no one really knows, but it kind of fits.

As for Catherine, she understands her heritage to be French for many generations back, as far as anyone has been able to trace genealogy. I get scolded when I suggest that the family tree may be particularly branchless. The report from 23andMe (shown in Figure 1.2) concurs, sort of. It shows French heritage back into the 19th century, as well as British and Irish ancestors during the same time period. This latter characteristic may be explained by Catherine's maternal antecedents being from Brittany, a maritime region in the extreme west of France. The language and culture there are Celtic and distinct from those of mainstream France—a distinction that was even more pronounced in the past. The Bretons have a lot in common with other Celtic people, including the Irish, Welsh, Manx, and Cornish. A number of her ancestors were sailors, which probably contributed to the genetic intermixing.

It's all interesting in a bourgeois sort of way, in which people living comfortably enough to spend a hundred bucks on a saliva test can look back at their ancestors, whose lives were foreign (in time, if not always location) and therefore conventionally romantic and interesting. In the case of people who find that their ancestors migrated from place to place, it becomes possible to conclude that the lives of the ancestors were unstable or otherwise crappy enough to motivate them to relocate, which allows the patrons of 23andMe to burnish their claims to modest origins. It's all good fun and harmless enough, facilitating European stuff on North American Christmas trees and a bottomless market for generic Irish music among Boomers in the former colonies.

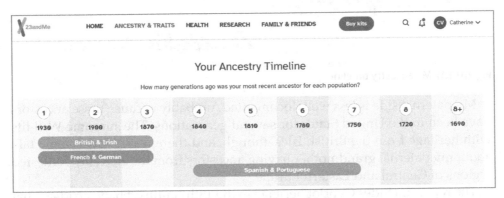

Figure 1.2: My wife's ancestry timeline

There's a dark side to genetic ancestry services like that of 23andMe, though. Some people assign extreme importance to their personal ancestry and believe that certain types of genetic heritage are inherently better than others. A scan of white supremacist websites, for example, will reveal screen shots of ancestry reports from 23andMe and similar services, posted by people attempting to fit in with the group. It's simultaneously risible and sad—or would be so, if it weren't for the undercurrent of violence—to see genetic profiling used in this way. Genetic testing is like any other power tool, though, capable of being used for ill as well as for good.

Perhaps the best part of 23andMe is the fact that it makes available what it calls *raw data*, which is a text file of 7–10 MB containing information about base pair sequences (see Figure 1.3).

You can take the 23andMe raw data file and use it as input to many of the other services listed earlier in this chapter.

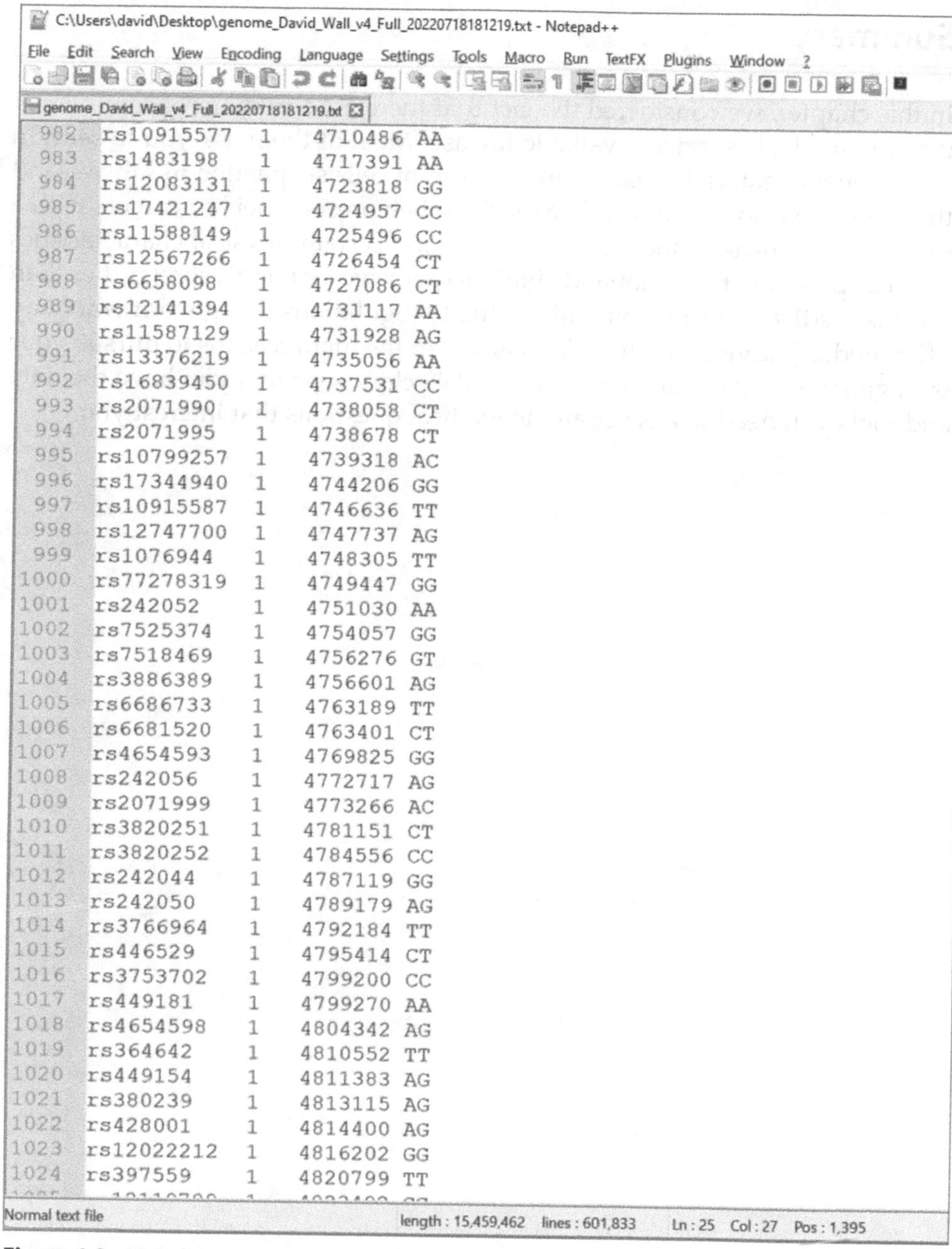

rs10915577	1	4710486	AA
rs1483198	1	4717391	AA
rs12083131	1	4723818	GG
rs17421247	1	4724957	CC
rs11588149	1	4725496	CC
rs12567266	1	4726454	CT
rs6658098	1	4727086	CT
rs12141394	1	4731117	AA
rs11587129	1	4731928	AG
rs13376219	1	4735056	AA
rs16839450	1	4737531	CC
rs2071990	1	4738058	CT
rs2071995	1	4738678	CT
rs10799257	1	4739318	AC
rs17344940	1	4744206	GG
rs10915587	1	4746636	TT
rs12747700	1	4747737	AG
rs1076944	1	4748305	TT
rs77278319	1	4749447	GG
rs242052	1	4751030	AA
rs7525374	1	4754057	GG
rs7518469	1	4756276	GT
rs3886389	1	4756601	AG
rs6686733	1	4763189	TT
rs6681520	1	4763401	CT
rs4654593	1	4769825	GG
rs242056	1	4772717	AG
rs2071999	1	4773266	AC
rs3820251	1	4781151	CT
rs3820252	1	4784556	CC
rs242044	1	4787119	GG
rs242050	1	4789179	AG
rs3766964	1	4792184	TT
rs446529	1	4795414	CT
rs3753702	1	4799200	CC
rs449181	1	4799270	AA
rs4654598	1	4804342	AG
rs364642	1	4810552	TT
rs449154	1	4811383	AG
rs380239	1	4813115	AG
rs428001	1	4814400	AG
rs12022212	1	4816202	GG
rs397559	1	4820799	TT

Figure 1.3: 23andMe raw data

Summary

In this chapter, we considered the fact that there are a number of commercial genome-analysis services available for use. Most of them ask you to send in some genetic material—usually in the form of cells suspended in saliva—and then test for various markers. The services then present their customers with a report, which usually focuses on ancestry and certain physical characteristics (such as probably hair color and ability to perceive certain flavors) rather than serious medical information. This is due to regulations that govern who may offer medical advice, and how. This is one of the main reasons to pursue your own genome sequencing work on the AWS cloud—so that you have the data and tools you need to answer all the medical questions that interest you.

# CHAPTER 2

A Crash Course in Molecular Biology

To properly analyze genomes, you need to understand some key biology concepts. Biology is a vast field, and this chapter focuses on areas that are fundamental for extracting meaningful information from genomes.

DNA

As you are probably aware, your genome encodes all the information necessary to build and maintain your body. It is composed of a series of about three billion letters, which can be either A, G, C, or T, and is stored inside the nucleus of each of your three trillion nucleated cells. Red blood cells do not have a nucleus and therefore are your only cells without a copy of your genome.

Note that all living organisms—such as animals, plants, fungi, or bacteria—also have a genome, composed of the same letters A, G, C and T. This book focuses on the analysis of the human genome, but all concepts and techniques are applicable to other species. You can apply them to the analysis of the genome of your cat, goldfish, potted plant, or slime mold growing in your front lawn. In genomics, it is sobering to see that humans are just a life species among many others.

Physically speaking, your genome is encoded into chains of repeating sub-units called *nucleotides*. Each nucleotide consists of a sugar group, a phosphate group, and a base that can be either adenine (A), guanine (G), cytosine (C), or thymine (T). The sugar group is called *deoxyribose*; hence, these long molecules are named *deoxyribonucleic acid (DNA)*. A chain of nucleotides is referred to as a *DNA strand*. To make this chemical structure more stable against mutations, your genome is actually stored as two complementary DNA strands. The second strand does not provide any more information and is simply a copy of the first strand, where A has been replaced by T, G by C, T by A, and C by G. The two DNA strands bind via weak chemical bonds called *hydrogen bonds* (the bases A pair with T, the bases G pair with C), coil around each other, and form the elegant and well-known DNA helix structure.

DNA is a convenient and compact way of storing information. It has already been used to store small, compressed MPEG movies and data of up to 200 MB.

In living organisms, the long DNA molecule is broken into segments called *chromosomes*. Species have various numbers of chromosomes: the fruit fly has 8 chromosomes, spinach and slime mold have 12, humans have 46, dogs and chickens have 78, and carp have 100. Bacteria stand apart as they have a single chromosome of circular shape, which is free floating as bacteria do not have a nucleus. The 46 human chromosomes are divided into two sets of 23 chromosomes, with each set inherited from each parent. Our chromosomes are numbered from 1 to 22, with the last two chromosomes, X and Y, determining our sex.

Not to be forgotten, each of our cells also contains between 80 and 2,000 small oval structures called *mitochondria* that produce the energy required for the cell. Mitochondria are thought to be bacteria that have been internalized by the cells of our remote primitive ancestors some 1.45 billion years ago. Each of our mitochondria has its own DNA, which has a circular shape (like in bacteria), unlike our linear chromosomes. We inherit our mitochondrial DNA exclusively from our mother. This allows genealogical researchers to study our maternal lineage many generations ago.

The total length of our genome, if we put our chromosomes end to end, would be nearly 2 meters. To fit into the tiny nucleus of each of our cells, our DNA is first wound around proteins called *histones* and then tightly folded and coiled. This packing complex of DNA and histone is called *chromatin*. When the chromatin is tightly packed (also called *condensed*), the DNA cannot be accessed. When a region of DNA needs to be read, the chromatin is unpacked and is said to be *open*.

Now, let's look at some chemical nuts and bolts to clarify an important concept that is sometimes confusing when interpreting genomics databases. DNA strands have an orientation: at one end of the strand, the base is connected to the fifth carbon atom of the sugar, whereas at the other end, the base is attached to the third carbon. Each end of a DNA strand is therefore nicknamed 5′ or 3′. By convention, DNA sequences are represented from 5′ to 3′. Because the second complementary strand does not represent any additional information, usually only the first strand (called the *coding strand* or *Plus strand* or + *strand*) is represented, whereas the complementary strand (*Minus strand* or − *strand*) is omitted (see Figure 2.1). Note that the Minus strand, read from 5′ to 3′ as per the convention, is the reverse complement of the Plus strand. Each paired group of letters of the double-stranded DNA molecule is referred to as a *base pair* (bp). So, the sequence shown in Figure 2.1 consists of 25 bp. Confusingly, though genomics databases describe DNA mutations or variants on the Plus strand, they are sometimes described on the Minus strand. It is necessary in that case to transform the nucleotide information into its complement (A to T, G to C, and vice versa).

Actual DNA Molecule	Common Representation
− strand: 5′ ATCCGCAATTACCGCAATGTGAATC 3′ llllllllllllllllllllllllll + strand: 3′ TAGGCGTTAATGGCGTTACACTTAG 5′	5′ ATCCGCAATTACCGCAATGTGAATC 3′

Figure 2.1: DNA molecule representation

Crucially, as the principle of life revolves around replicating itself, DNA can easily be replicated via convenient cell enzymes called *DNA polymerases*. This enables cells to receive an identical copy of the genome after each cell division. In practice, the chromatin of the chromosomes is decondensed, the two strands of their double helix are unwound, and the DNA polymerase synthesizes a new complementary DNA strand for each of the two initial strands. At the end of the process, the DNA is fully duplicated into identical double helices. The chromosomes then align in the center of the nucleus. Fibers attach to the center (called *centrosome*) of the chromosomes and pull one copy of each chromosome to opposite poles of the cell. This process by which two new nuclei are formed, in view of creating two identical daughter cells, is caused *mitosis*. Finally, the cell itself divides into two cells with each an identical genome.

How long does it take for a cell to replicate? Bacteria have very small genomes that can be replicated quickly, as well as short growth phases; in consequence, bacteria can divide in up to 15 minutes. In contrast, fast-growing human cells take about 24 hours to divide. The cell starts with a growth phase (called *G1*) of about 11 hours, where the cell secretes enzymes and nutrients that will be required for the cell division. The cell then undergoes a phase of DNA synthesis (called *S phase*) to duplicate the genome, which takes about 8 hours. It is followed by a second growth phase (called *G2*) of about 4 hours. Finally, the nucleus division (mitosis) and cell division can occur. This last phase, called *M phase*, is the most visual and spectacular and takes only one hour.

The chromosomes are not usually visible under the microscope, as they are all bundled together in the nucleus. During mitosis, the chromosomes are condensed and neatly align in the center of the cell. It is during this phase that the chromosomes can be seen and photographed under a light microscope when stained with a dye. This technique is called *karyotyping*. Figure 2.2 shows a normal human karyotype. Chromosomes have been reordered from chromosome 1 to 22 and the X sex chromosome (also called *allosomes*). You can tell it is from a male, because there is an X chromosome and a Y chromosome (not paired, unlike chromosomes 1–22).

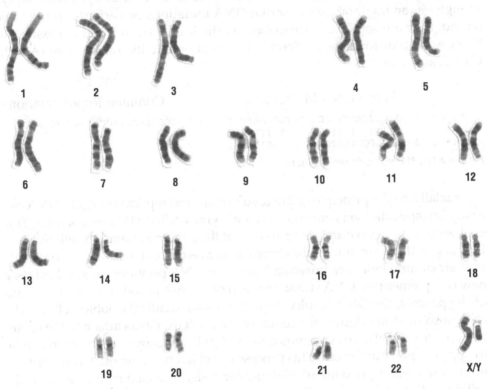

Figure 2.2: Human karyotype

"National Human Genome Research Institute / Wikimedia Commons / Public Domain"

Karyotypes can identify abnormalities such as an extra copy of chromosome 21, which causes Down syndrome. Trisomies of other autosomal chromosomes are rarely compatible with survival beyond birth or a few months, and most often cause miscarriages. An extra copy of sexual chromosomes is more common (more than 3 per 1,000 births). Males can carry an extra X chromosome, resulting in a karyotype 47, XXY, which causes the Klinefelter syndrome (infertility and small testicles), or an extra Y chromosome (karyotype 47, XYY) associated with a tall stature and an increased risk of learning difficulties. Females can have an extra X chromosome (karyotype 47, XXX), though this usually does not lead to any particular symptom.

On the karyotype, chromosomes appear as long thin cylinders with a thin waist in the middle. This thinned region, called the *centromere*, corresponds to the area where the chromosome was attached to the fibers in order to be pulled apart to the poles of the cell during mitosis. The ends of each chromosome contain a chain of repetitive letters. For humans and other vertebrates, the sequence is AGGGTT (TCCCAA on the complementary strand), repeated about 2,500 times in humans. These are *telomeres*, and their role is to protect the ends of our chromosomes from damage or from fusing with other chromosomes. Limitations in the DNA replication machinery cause telomeres to shorten at each cell division. In addition, oxidative stress and free radicals can further shorten telomeres. When telomeres are too short, cells stop dividing and enter a state called *senescence*. Senescent cells send inflammatory signals and are associated with aging and disease.

The upper half of each chromosome, which is shorter, is called the *p arm* (*p* stands for "petit," French for small). The lower, or longer, part of the chromosome is named the *q arm*. The chromosome arms have been further conventionally divided into bands and sub-bands. For example, 13p11.2 refers to chromosome 13, arm p, band 11, sub-band 2. On genome browsers, chromosomes are represented horizontally, with the p arm conventionally placed on the left.

How does our genome evolve during our lifetime? At our birth, all our cells have the same genome. The exception concerns relatively rare cases of *mosaicism*, where an individual has the same mutation only in a certain part of the body, often due to the mutation occurring at an early embryonal stage in a subset of their cells. So, in most cases, all our cells at birth have the same genome, which is the replica of the genome of the initial fertilized egg produced by our biological parents. Because the egg and sperm that generated our first cell are called *germ cells*, our birth genome is called our *germline genome*. As we age, we will progressively accumulate mutations, such as UV-induced mutations in our skin, pollutant-induced mutations in our lungs, or cancers. The genomes of these tissues will have significantly drifted from the original germline genome and are called *somatic genomes*. The genome analysis methods slightly differ between germline and somatic genomes, so it is important to understand the difference.

DNA at Work: RNA and Proteins

DNA is a repository of information and does not do anything by itself. But it can be read and acts as a template to produce RNA and proteins, which are the workhorses in our bodies. The production of RNA from DNA is straightforward and is called *transcription*, whereas the generation of protein from RNA is more complex and is called *translation*. We need at this stage to introduce the notion of gene, which is a string of letters in our genome that codes for a functional unit.

How many genes do we have? The question is hotly debated and depends on the definition of a gene. The definition of a gene used to be based on the central dogma of molecular biology: the DNA of a gene is the template for producing RNA, which is then translated into protein. We have about 20,000 protein-coding genes, down from a prior evaluation of about 30,000 before the completion of the sequencing of the first human genome. However, we now know that our genome is widely transcribed and includes more than 15,000 nonprotein coding genes, which are genes that produce RNA, which is never translated (called *non-coding RNA*, or *ncRNA*). The function of many of these ncRNAs is unknown. Some ncRNAs help translate RNA into proteins, while many others regulate the transcription of genes.

From a nomenclature perspective, genes are named by a governing body called the *HUGO Gene Nomenclature Committee*. HUGO stands for HUman Genome Organization. In practice, genes are referred to in the scientific literature with many different names, which can be extremely confusing and complicates bioinformatics programs. The Genecard database (www.genecards.org) keeps track of previous names and provides excellent gene-centric integrated data integrated from many web sources.

It came as a surprise, and a bit of a blow to our ego, that humans have only about 20,000 protein-coding genes. We compensate this small number by adding complexity to our proteins though a process called *alternate splicing*. Our protein-coding gene sequences are composed of a succession of coding regions (called *exons*) and noncoding regions (called *introns*), which are excised (or spliced out) from the final RNA and never make it into protein. The same gene can produce different proteins (called *isoforms*) by splicing out different exons as well. If a gene is composed of four exons, one isoform can have all four exons (exon1-exon2-exon3-exon4), while another might skip the third exon (exon1-exon2-exon4).

The transcription of DNA into RNA is performed by the enzyme RNA polymerase. DNA is always read from 3′ to 5′, and synthesized from 5′ to 3′. If the gene is in the Plus strand, the enzyme transcribes the Minus strand, producing a complementary sequence that is identical to the original gene DNA sequence (see Figure 2.3). There are, however, two differences: thymine (T) is replaced by uracil (U), and the sugar is a ribose—the reason for the name ribonucleic acid. Note that RNA is composed of a single strand.

If the RNA is to be translated into a protein, it is further processed into a *messenger RNA* (*mRNA*). First, a chemical group is added at its 3′ end and a polyA sequence at its 5′ end. Second, the introns, and possibly some exons (alternate splicing), are spliced out. The mRNA leaves the nucleus of the cell and is translated into a protein by a complex machinery of noncoding RNAs and proteins called the *ribosome*. The mRNA is read three letters at a time.

Each group of three letters is called a *codon*. Each codon corresponds to a single amino acid. A protein is a long chain of amino acids. The code for transforming codons into amino acids is called the *genetic code* and is valid for most living organisms, including bacteria. It means that a human gene sequence can be inserted into a bacterial genome and produce the same protein as if it originated from a human being. This principle is widely used to industrially produce proteins, called *recombinant proteins*, such as insulin for diabetics.

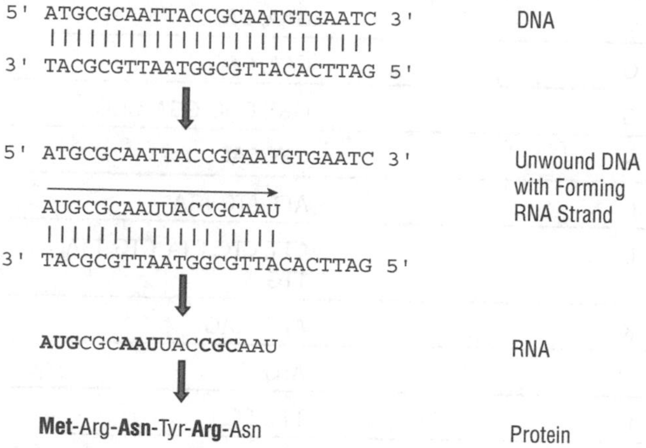

```
5'  ATGCGCAATTACCGCAATGTGAATC  3'          DNA
    ||||||||||||||||||||||||||
3'  TACGCGTTAATGGCGTTACACTTAG  5'

                    ↓

5'  ATGCGCAATTACCGCAATGTGAATC  3'          Unwound DNA
    ─────────────────────────→              with Forming
    AUGCGCAAUUACCGCAAU                       RNA Strand
    |||||||||||||||||||
3'  TACGCGTTAATGGCGTTACACTTAG  5'

                    ↓

    AUGCGCAAUUACCGCAAU                       RNA

                    ↓

    Met-Arg-Asn-Tyr-Arg-Asn                  Protein
```

Figure 2.3: Transcription and translation

Figure 2.4 shows the genetic code. For convenience, we have used T instead of U in the codons, as it makes it easier to interpret in the context of genome analysis. There are 20 possible amino acids in a protein. Amino acid names are abbreviated in databases—for example, methionine can be abbreviated to Met or M. The code is said to be *degenerate*, as several codons correspond to the same amino acid. If you look carefully at the table, you will notice that the redundancy often comes from the third letter of the codon. The first amino acid of a protein is always methionine, and the stop-codon signals the end of the protein. In Figure 2.3, the first three letters, which form the first codon, of the sequence to translate are AUG, which corresponds to the amino acid methionine (Met). The second codon is CGC, which corresponds to arginine (Arg). The third codon is AAU, which corresponds to asparagine (Asn). Each amino acid is joined to the previous one via a chemical bond called a *peptide bond*, until all the codons of the mRNA sequence have been translated. The final protein, composed of this chain of amino acids, is then complete.

AMINO ACID	SLC	DNA CODONS
Alanine	A	GCT, GCC, GCA, GCG
Arginine	R	CGT, CGC, CGA, CGG, AGA, AGG
Asparagine	N	AAT, AAC
Aspartic acid	D	GAT, GAC
Cysteine	C	TGT, TGC
Glutamic acid	E	GAA, GAG
Glutamine	Q	CAA, CAG
Glycine	G	GGT, GGC, GGA, GGG
Histidine	H	CAT, CAC
Isoleucine	I	ATT, ATC, ATA
Leucine	L	CTT, CTC, CTA, CTG, TTA, TTG
Lysine	K	AAA, AAG
Methionine	M	ATG
Phenylalanine	F	TTT, TTC
Proline	P	CCT, CCC, CCA, CCG
Serine	S	TCT, TCC, TCA, TCG, AGT, AGC
Stop codons	Stop	TAA, TAG, TGA
Threonine	T	ACT, ACC, ACA, ACG
Tryptophan	W	TGG
Tyrosine	Y	TAT, TAC
Valine	V	GTT, GTC, GTA, GTG

Figure 2.4: The genetic code

Proteins perform broad functions in the body. They can be structural and shape the structure of our tissues and cells, like the myosins and related proteins that form our muscle, or the collagen, elastin, and keratin that are embedded in our tissues. They can be small astounding nanoscale motors. The protein ATP synthase, for example, has a rotary engine; its rotor spins around and uses this mechanical energy to produce ATP, which is the energy currency of the cell. The kinesin proteins form the transport system of the cell: they attach to a cargo of compounds that need to be transported and drag this cargo to another location in the cell by walking along fibers called *microtubules*. Kinesins have two small globular structures that look like two small feet and literally walk along the microtubules, at a rhythm of about 100 steps per second. Some proteins act as storage, such as ferritin, which stores iron, or transporters, such as hemoglobin, which transports oxygen. Proteins can act as chemical messengers, such as the interleukins that send messages to our immune system. Proteins can also form receptors, channels, and pumps on the surface of our cells. Finally, many of our proteins are enzymes that catalyze the numerous chemical reactions in our bodies. Proteins often do not work in isolation and assemble into large complexes.

Unlike DNA, which has a simple structure, proteins fold into complex shapes that are determined by their sequence. We do not well understand the rules that govern the correspondence between protein sequence and protein shape. Protein folding is actually a great scientific conundrum. To complicate matters, proteins are not static and can change shape to carry out the work they are supposed to do. These varying shapes are called *conformational changes*. The shape or structure of about 200,000 proteins has been determined experimentally, mainly via X-ray crystallography and electron microscopy. Most proteins are represented in a single conformation, but some are done with several conformations. All 200,000 protein structures are publicly available in the Protein Data Bank (www .rcsb.org), where they can be visualized with 3D simulations. The best protein shape prediction program is an artificial intelligence system called Alphafold, which consists of neural networks trained over the protein structures of the Protein Data Bank. Alphafold recently released predicted structures for all human proteins, with a high confidence level for 58% of amino acids. All these computed predictions are available on the Protein Databank web site.

What exactly is an amino acid? It is a small molecule composed of an amino group (chemical formula $-NH_2$, where N stands for nitrogen and H for hydrogen), a carboxyl group (chemical formula -COOH, where C stands for carbon and O for oxygen), and a side chain that is different for each of the 20 amino acids. The side chain will give the amino acid its specific properties, such as size, electrical charge, or faculty to attract or repulse water. Note that amino acids are sometimes called *residues*. Each protein begins with the first amino group (NH_2) of its first amino acid and ends with the carboxyl group (COOH) of its last amino acid. So, the beginning of a protein is called the *N-terminal*, or *amino-terminal*, and the end of a protein is termed the *C-terminal*, or *carboxyl-terminal*.

Proteins are said to have four levels of structure.

- The primary structure is the sequence itself.
- The secondary structure corresponds to recurring structural motifs.
- The tertiary structure is the overall shape of the protein.
- The quaternary structure is the assembly of the protein with other proteins to form larger multiprotein complexes.

The main secondary structure motifs are helices, where regions of the protein coil into a perfect helix (the most common type of helix turns every 3.6 amino acids in a clockwise manner), and beta-sheets, where nonconsecutive regions of the protein called *beta-strands* align together and form a kind of pleated sheet. These secondary motifs are created by weak bonds called *hydrogen bonds* (the same bonds that enabled the formation of the DNA helix). Most proteins are built from combinations of helices and beta-strands, which are interconnected by loop regions of variable length and irregular shape. Secondary motifs can form larger structures such as barrels and propeller-like structures. Beside the structural proteins and the receptor proteins located on the membrane of cells, the most common proteins have a globular shape, with a hydrophobic (fearing water) interior.

The cysteine amino acid deserves a special mention in protein folding, as it contains a sulfur atom. Two cysteine amino acids located in different regions of the protein chain, but adjacent in the three-dimensional structure of the protein, can oxidize and form a covalent bond called *disulfide bridge*. These bonds are important and can stabilize the structure of the protein. As an example, perming your hair at the hairdresser involves breaking the disulfide bonds of your hair and re-forming them when your hair is in the desired conformation.

Figure 2.5 displays the protein alcohol dehydrogenase, an enzyme that breaks down in the liver the alcohol we might drink. It actually converts alcohol into acetaldehyde, which is even more toxic, before it is converted into harmless acetate by another enzyme. There are actually nine different isoforms of alcohol dehydrogenase (reminder: isoforms are created by alternate splicing), and Figure 2.5 displays one of these isoforms. Proteins can be represented by different styles: on the left, the protein is drawn using a cartoon style. The helices are shaded dark, and the beta-sheets lighter. Visible in the center is a molecule of alcohol. On the side, the protein is represented using a small sphere for each atom.

Our DNA has many proofreading and repair mechanisms. However, not all DNA mutations can be repaired. If the mutation affects one of the exons of a protein-coding gene, the protein might be affected. Let's look at the possible scenarios. If the mutation corresponds to the third letter of a codon, the mutation might be silent, as the codon might still code for the same amino acid. If the mutation changes the amino acid, however, the impact will depend on whether the new amino acid has the same characteristics. Important characteristics include the size of the amino acid (glycine is the smallest, and tryptophan the largest), its electrical charge (positive, negative, or neutral), and whether it attracts water (hydrophilic) or fears water (hydrophobic). We will discuss tools in Chapter 11 that help evaluate the impact of these parameters on the shape and function of the protein. What are the worse mutations? If the mutation transforms the codon into a stop codon, the protein will be prematurely truncated. If the few base pairs at the border of the exon-intron are mutated, the splicing will be affected, and exons might be skipped erroneously. The last type of highly deleterious mutation is a frameshift, which occurs when an insertion or deletion ("indel") of one or two base pairs occurs; the translation by triplets of letters gets shifted by one or two positions, and all amino acids from that indel position become incorrect.

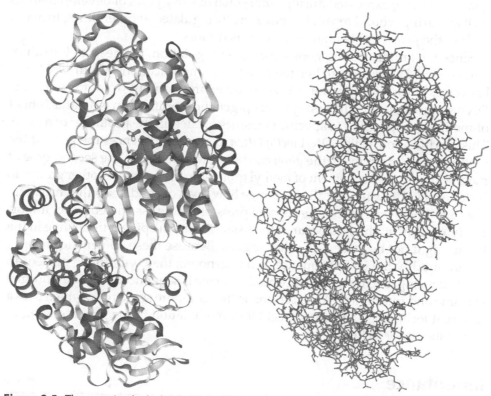

Figure 2.5: The protein alcohol dehydrogenase

What about mutations of ncRNA? Mutations of ncRNA that form part of the translation machinery, such as transfer RNAs or ribosomal RNAs, will perturb translation. What about mutations outside genes? The transcription of genes (also called the *expression* of genes) is regulated by proteins named *transcription factors* (TFs). TFs bind regions upstream of genes called *promoters* and *enhancers*. Enhancers can be quite far away in sequence from the gene they regulate. The blue eye color is thought to be caused by the mutation of an enhancer of the gene OCA2. This enhancer is located more than 20,000 base pairs from OCA2 and actually falls within the intron of another gene. This shows the complexity of evaluating the impact of mutations. Promoters and enhancers are disseminated throughout the genome, including the noncoding regions, and affect the way our genes are expressed.

We cannot emphasize enough the importance of the regulation of the expression of our genes. All our cells carry the same genome, but they perform widely different functions. The insulin gene INS needs to be expressed in the islet cells of the pancreas but should stay silent in most of your other cells. Regulation needs to occur not only by location and type of cells, but also in time: growth-promoting genes should be expressed during embryonal or early years of development, as well as during wound repair, but be carefully regulated at other times, to avoid fueling the growth of benign or cancerous tumors.

Since we are discussing gene expression regulation, we need to introduce epigenetics. *Epigenetics* are heritable chemical modifications of our DNA and histones. These modifications affect gene expression but do not modify the DNA sequence itself. The first type of epigenetic modification is an attachment of methyl group (-CH$_3$) to specific cytosines located in the promoter of a gene. This modification, called *DNA methylation*, completely silences the gene. If the methyl group is removed, the gene can be expressed again. The second type of modification is the attachment of methyl groups, acetyl groups, or other groups to histone tails. These modifications change how open or condensed the chromatin is and therefore alter the patterns of expression of the associated gene. Importantly, epigenetic changes occur at the tissue level. An epigenetic change in the brain might not be present in other organs. Because these epigenetic changes do not modify the DNA sequence of our genome, they will not be picked up by sequencing machines. A methylated cytosine is sequenced just like a normal cytosine and returns a C. An exception is the nanopore sequencer, which uses a different technology and can detect DNA methylation in the genome without extra laboratory techniques.

Inheritance

As we have seen earlier, we have two sets of chromosomes, with one set inherited from each of our biological parents. Germ cells in ovaries or testicles undergo a special type of cell division called *meiosis*, which generates daughter cells (egg or sperm) that contain only one set of chromosomes. An important event occurs during meiosis: while the paired chromosomes are aligned in front of each other in the first phase of meiosis, they exchange sections of DNA between each other. This event, called *crossing-over*, occurs on average three to four times for each pair of chromosomes. It means that sperm or eggs rarely contain any integral original chromosome from any parent. They have instead chromosomes that have exchanged DNA material between parents.

At egg fertilization, the egg and sperm fuse their nuclei together, as they prepare for the cell replication. The fertilized egg now contains the full two sets of chromosomes. The sperm mitochondria get progressively eliminated, and only the egg mitochondria, therefore maternal mitochondria, remain.

As we have received a set of chromosomes from each parent, we dispose of two copies for most genes: a maternal copy and a paternal copy. Genes can have several alternate forms, called *alleles*. For example, a particular position (locus) in the human genome may have two alternative nucleotides, or alleles, C or T. In genome databases, the term *reference allele* refers to the nucleotide that is found in the reference human genome. As the reference is built from a limited number of individuals, it is not always the major allele found in the population.

Single nucleotide variations are often called *single nucleotide polymorphisms* (SNPs, pronounced "snips"). SNPs traditionally referred to variants present in more than 5 percent of the population in order to exclude rare mutations. In practice, databases now list SNPs of extremely low frequency. We and many others prefer to use the generic term *single nucleotide variation* (SNV, pronounced "sniv"), which does not make any assumption on the frequency of the variation.

Our genome constitutes our genotype, which is all our genetic information. Our genotype has an effect on our phenotype, which consists of our observable physical properties, such as height, hair color, or behavior. If we inherit the same allele or SNV from both our parents, we are homozygous for that SNV. If we inherit different SNVs (for example, an A from one parent and a T from the other parent), we are heterozygous for that SNV.

In the genome, combinations of alleles at positions close to each other on the same chromosome tend to happen more often than would be expected by pure chance; this suggests they were inherited together. These combinations of alleles are called *haplotypes*. The technique of linkage disequilibrium quantifies how often two alleles or sequence variants are inherited together; alleles that are always inherited together are said to be in *linkage disequilibrium*.

Alleles or SNVs can have several modes of inheritance. If the allele or SNV is associated with a strong phenotype, it is worthwhile doing a family tree indicating which family members are affected. This tree is called a *family pedigree*. Figure 2.6 illustrates an example of family pedigree. Males and females are respectively represented as squares and circles. Affected family members have black-filled shapes, unaffected carriers have half-filled shapes, and deceased family members are unceremoniously crossed out. The "pro-band" corresponds to the affected family member that triggered the investigation.

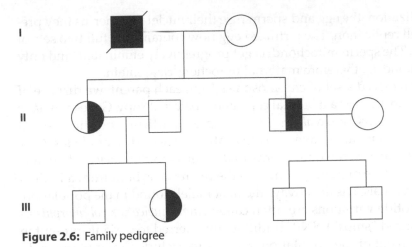

Figure 2.6: Family pedigree

The following are the main modes of inheritance:

- **Autosomal dominant**: A single SNV inherited from one parent can cause the condition, and the normal version inherited from the other parent cannot compensate for it. On the family pedigree: affected individuals usually have an affected parent. The SNV-associated phenotype occurs in every generation. Familial hypercholesterolemia is an example of autosomal dominant condition.

- **Autosomal recessive**: The SNV needs to be inherited from both parents to cause the condition. On the family pedigree, both parents of the affected individuals are carriers. It's a condition typically not seen in every generation. Cystic fibrosis and sickle cell anemia are examples of autosomal recessive conditions.

- **X-linked dominant**: The SNV is located on the X chromosome, and one copy of the SNV is enough to cause the condition. So, men and women can be equally affected. Affected men pass the condition to all their daughters but to none of their sons.

- **X-linked recessive**: The SNV is located on the X chromosome, and the presence of a normal version of the gene can rescue the phenotype. Males have only one X chromosome and so will be affected if they have the SNV. Females have two X chromosomes so are less likely to be affected. Hemophilia A is one example.

- **Y-linked SNV on the Y chromosome**: This condition is passed from father to son.

- **Co-dominant**: The gene can have two types of SNV or allele. Both alleles influence the phenotype. The ABO blood group is an example.

- **Mitochondrial**: Mitochondria are inherited only from the mother. SNVs located in mitochondria can cause the condition in men and women. Only women, however, will transmit the condition to their progeny.

The preceding modes of inheritance apply to conditions caused by a single gene. Many conditions are caused by several genes and are said to be *polygenic*. Complex traits or conditions, such as height, diabetes, or obesity, are influenced by hundreds of loci in our genomes. To identify these loci, genome-wide associated studies (GWAS) are conducted, where the genomes of a large number of individuals with the trait or complex disease are compared the genomes of healthy controls. You will see in Chapter 10 how to best use the findings of GWAS studies.

Summary

We hope you have gained an understanding of genomics from this chapter. We explained how the genetic code contained in DNA translates to amino acids, which in turn build proteins and the larger structures that make up living things.

Next, we will show you how to take a blood sample from yourself or someone else whose genome you want to study and how to use a service laboratory to get computer data derived from the biological sample.

Obtaining Your Genome

Unless you have access to a lot of expensive laboratory equipment and possess the skills to use it, the process of obtaining your genome will involve engaging a laboratory to take some of your biological material, extract the nuclear DNA from it, and sequence that DNA to yield a data file containing your genome.

Preparing to Have Your Genome Sequenced

As you think about having your genome sequenced, there are some things you need to know before you start.

Can It Affect My Insurance?

In the United States, the Genetic Information Nondiscrimination Act (GINA) makes it illegal for health insurance providers in the United States to use genetic information to deny health coverage or increase premiums. (Full details and exceptions are at ghr.nlm.nih.gov/primer/dtcgenetictesting/dtcinsurancerisk.) A similar law is in vigor in Australia for Medicare and private health funds. In both countries, however, genetic information can be used to increase premiums for other types of insurance, such as life insurance or disability insurance.

Most countries of the European Union (`www.iuscommune.eu/html/activities/2013/2013-11-28/workshop9a_Goessens.pdf`) prohibit the use of genetic information for increasing premiums or denying coverage for all forms of insurances: health insurance, life insurance, etc.

Privacy

Before you sign with a company that provides genome sequencing, read carefully the fine print in terms of confidentiality of produced genomic data. Some companies provide low prices as they sell some information from the de-identified genomes to pharmaceutical companies or other research institutes. This is not bad in itself and can contribute to better cures, but it needs then to be transparently agreed on both sides. Some more expensive providers provide complete genomic data confidentiality.

Humility and Levelheadedness

The more we explore the human genome, the more we discover new levels of complexity: 16 years after the completion of the first human genome, we still do not understand the function of half of the human genes. The "junk DNA" is not considered junk anymore; it is littered with regulatory regions and is widely transcribed, whereas it was previously thought to be silent. New complex levels of regulation of the genome are discovered every few months. Scientists still widely disagree on the overall proportion of the genome that is functional. (The range varies between 15 percent and 80 percent.)

In the past, sequencing efforts focused on the sick. Now that more and more healthy people are being sequenced, we discover that mutations that were thought to be very deleterious are found in individuals living long healthy lives. Our genomes and our cells have a lot of redundant mechanisms that help them cope with mutations. With the exception of Huntington's disease, which is a catastrophic disease that can be predicted with high accuracy from specialized genetic testing, most mutations or gene variants can only produce an increased risk and require additional mutations (genetic background) and environmental factors to actually produce disease. In other words, relax. Even the very well-known APOE4 gene variant, associated with Alzheimer's disease, only moderately increases your risk of getting the disease. Note anyway that Huntington disease, as well as other related polyglutamine disorders caused by the expansion of repeats in the genome, is hard to detect via whole-genome sequencing and is always diagnosed instead via specialized genetic tests.

Validation with a Clinically Accredited Test

In the unlikely event you make any finding that concerns you and would prompt you to take medical action, always validate with a clinically accredited test, and consult with a genetic counselor. A clinically accredited test will have more quality controls in the lab, ensuring, for example, that there has not been any swapping of samples. The knowledge you will have accumulated by analyzing your genome and researching your mutations will enable you to have an active role in your discussions with your genetic counselor and medical professionals and will help you decide on the best course of action.

Alternatives to Using Your Own Genome

If you are not sure yet if you would like to sequence your own genome, or the genome of a relative, you can start instead with a publicly available genome.

International Genome Sample Resource (IGSR)   IGSR (`www.internation algenome.org/`) provides access to the data generated by the 1000 Genomes project and expands to new data sets. IGSR includes the whole-genome sequence (WGS) at low depth of about 3,000 individuals and includes the WGS at high depth of a subselection of individuals. The data is also stored in an Amazon S3 bucket (`s3://1000genomes`).

Genome in a Bottle (GIAB)   GIAB (`jimb.stanford.edu/giab`) is a consortium that develops reference standards, methods, and data to help translate whole-genome sequencing into clinical practice. GIAB sequenced at high depth and on several technological platforms the whole genome of an American female of Utah, of European ancestry, referred to as *NA12878*. This woman's genome is probably the most studied genome ever and is used as a benchmark for testing many technological methods. GIAB also similarly studied two trios (mother-father-child) of Ashkenazi Jewish and Han Chinese ancestries.

Specifying Lab Work

To engage a lab, you have to specify what you want done. This section describes the options you have.

Depth

The cost of WGS is highly dependent on the requested depth of sequencing. The *depth*, or coverage, corresponds to the number of times each given nucleotide has been sequenced. It is usually recommended to sequence at a depth of 30 times (30×) for germline genomes, which means the genome we were born with. This corresponds in theory to an average of 30 reads to determine the nature of each letter of your genome, or 2×15 reads if you are heterozygous at a particular position in your genome. In practice, the number of reads can turn out to be much lower: certain areas of the genome regions of the genome do not amplify well and have therefore a lower number of reads; bacterial contamination can also lower the effective depth of the sample, as bacterial DNA will be sequenced as well as human DNA. Choosing a low depth makes it difficult to distinguish true mutations from sequencing errors.

To determine which mutations we have accumulated since birth (somatic genomes), such as which mutations have developed in a cancerous tumor, it is recommended to sequence at a depth of 90×—or more, in case of sample degradation.

Sample Type

Saliva and buccal samples can be highly contaminated with bacteria, even if the container solvent is deemed antibacterial. In addition, some buccal cells can be mutated in case of frequent use of alcoholic mouth rinses or beverages. So, blood samples are best. Note that most of us develop progressive mutations in our white blood cells as we age, through a process called *clonal hematopoiesis*; these clonal cells usually stay rare until an advanced age and do not perturb whole-genome sequencing.

Type of Output Files

Whole-genome sequencing typically delivers quality-control files and a FASTQ file, which is a specially formatted text file that contains raw sequence information. Most sequencing companies also offer an optional bioinformatics analysis, which we recommend to take, as a control for your own analysis. The bioinformatics analysis also delivers an annotated variant call file (VCF)—another special text file that describes how the genome under analysis differs from standard reference genomes. Furthermore, there is an analysis of copy number variation (CNV), which is when there are multiple repeats of a sequence that occurs only once in reference genomes—a condition that is associated with certain pathologies. There will also be information about large regions of inversion and deletion (also known as *structural variations*) and general statistics.

Sequencing Technology

A number of competing sequencing technologies exist. These include the following:

Illumina and BGI Short Read Technologies   These massively parallel sequencing technologies are by far the most popular. They rely on shotgun sequencing, where the DNA is chopped in small fragments, which are individually sequenced. The first step is library preparation: the DNA is fragmented, the ends of the fragments are repaired, adapters (synthetic double stranded DNA sequences) are ligated to each end of the fragments, and the DNA fragments with adapters are selectively enriched. The sample is then loaded into a specialized glass slide called a *flowcell*, where the amplification and actual sequencing takes place. At the bottom of the flowcell are attached DNA sequences that are complementary to the adapters. The fragments introduced into the flowcell randomly bind by hybridization to these attached sequences. DNA polymerases are then entered into the flowcell to create about 1,000 copies of each fragment. The actual sequencing, called *sequencing by synthesis,* can now take place. Modified nucleotides are introduced into the flowcell. These specialized nucleotides force the DNA polymerase to add only one nucleotide at a time and have a fluorescent label. The DNA polymerases synthesize a complementary strand for each fragment, using the modified nucleotides. A camera records all the fluorescent emissions and determines from the fluorescence wavelength which base (A, T, G, or C) was added.

The sequencing can be done along one end of the fragment (*single-end sequencing*). By adding a few extra steps to the process described previously, it is possible to sequence both ends of the fragment (*paired-end sequencing*). Whether single-end or paired-end sequencing, the sequencing is typically done for a length of 50 to 150 bases, though the latest Illumina platform (NovaSeq) can go up to 250 bases. The human genome has many repetitive regions where it is difficult to properly align short sequences obtained from short read sequencing. To facilitate the alignment in repetitive regions, as well as to improve the detection of large structural variations in the genome, it is advisable to use paired-end sequencing, as well as reads as long as possible.

PacBio   This technique, from Pacific Biosciences, relies on single-molecule, real-time (SMRT) sequencing to allow for long reads of about 50,000 bases on average. The PacBio platform has a higher sequencing error rate than Illumina/BGI. It is still very expensive and is reserved for research.

Nanopore   The Nanopore approach, from Oxford Nanopore Technology, is totally different. The DNA molecule is fed through a small channel called a *nanopore*, which generates an electric current that can be interpreted to guess the nucleotide sequence. Nanopore can generate very long reads, and the device (Nanopore MinION) is about the size of a portable telephone. However, Nanopore has a high error rate, so it is usually used in complement to traditional short read technologies to work on areas of the genome with high repeats. If, for example, a mutation is suspected close to the centromeres or telomeres, or near mobile elements with high repeats, nanopore might help. It is also useful in the sequencing of the genomes of new species, such as bacteria.

Chromium 10x   Chromium 10x requires special library preparation, in which the reads are barcoded so that they can be linked later. The reads are then sequenced with Illumina. It is more expensive than the usual short read sequencing, but less expensive than PacBio and more accurate than Nanopore. Chromium 10x can be useful if the DNA to be sequenced is of good quality and therefore not too fragmented.

Genome vs. Exome vs. SNP Arrays

Whole-exome sequencing (WES) sequences only genes and splicing regions, so it misses a lot of key regulatory regions. Whole-genome sequencing (WGS) has no such problem, as it covers by definition the whole genome. Given the falling sequencing prices, WES is only marginally less expensive than WGS. Single nucleotide polymorphism (SNP) arrays sequence only a few million letters throughout the genome, so will miss a lot of variants. In addition, as SNP arrays rely on hybridization of DNA to nucleotides attached to a glass slide, many SNPs will not get a reading if the hybridization fails. (The SNP is reported in that case as a "no call.")

Engaging a Laboratory

A number of labs will do this kind of work. We chose to use BGI, headquartered in Shenzhen, China, which has been doing whole-genome sequencing for more than a decade and asks a reasonable price for its work. BGI is a large and reputable company—possibly the largest genomics company in China, with shares traded on the Shenzhen exchange and annual revenue in excess of USD $250 million. Despite that, the company seems completely happy to do business with individuals, unlike other genomics companies that focus on doing business with large hospitals and medical laboratories.

We wrote to BGI and agreed on a specification of work to be performed:

- DNA extraction, meaning the lab would take the blood sample and extract DNA from it. See Chapter 2, "A Crash Course in Molecular Biology," for more discussion of the actual process by which this is done.

- Human whole-genome sequencing, meaning the generation of FASTQ files containing the whole-genome sequence.

- Paired-end sequencing of 150 base pairs per end, meaning that each fragment will be sequenced from each end for a length of 150 base pairs. For each fragment, there will be two reads of 150 base pairs that start at each end and go toward the middle of the fragment. Depending on the size of the insert, the reads might overlap a bit in the middle, or there might be an unsequenced part in the middle, if the fragment is longer than 300 base pairs. Technically, the fragment includes *adapters* at each end, which are artificial strings of nucleotides that make it easier for sequencing equipment to process DNA strands. The natural DNA fragment without the adapters attached is called the *insert*.

The alternative to paired-end sequencing is single-end sequencing, which costs less. We decided to go with paired-end sequencing of 150 base pairs (so 2×150 bp for each fragment) in order, as mentioned previously, to be able to have better alignments in repetitive areas of the genome and to improve the detection of structural variants.

- The alignment information should go into Binary Alignment Map (BAM) files—machine-readable files that describe how sequences in a genome under study are aligned, relative to a reference genome—of approximately 135 GB for a depth of 45×. (There is more information on BAM files and the information they contain later in this chapter.)

In other words, the lab will extract DNA from the sample we provide and then perform WGS by reading fragments from each end. It will put those sequences in FASTQ files and align those sequences to the reference genome, with the output of that process going into BAM files.

The results would be sent to us on a portable hard drive via express courier and will cost just under USD $1,500. We had to pay before the company would send us the results.

Getting a Tissue Sample for DNA Extraction

The process of obtaining your genome begins with getting a sample of *you*. Almost any tissue will do, in theory, but the easiest and best way to get a sample of cells that contain nuclear DNA is to take a fresh blood sample, seal it in a sterile container, and ship it to the laboratory for DNA extraction and analysis. BGI gave us a specification of what it needed: a 5 mL sample of whole blood, preserved in a vacuum vial.

To address some common misconceptions about where DNA comes from:

- Hair does not contain nuclear DNA, though it does contain mitochondrial DNA (mtDNA). An organism's genome is defined by the DNA in the nuclei of its cells, so hair is not an acceptable source. Criminal forensics work in which a "DNA match" is established thanks to hair either is based on mtDNA or relies on nuclear DNA found in hair follicles.

- Finger- and toenails *do* contain nuclear DNA and so technically can be used as the basis for genome sequencing. However, because of the rugged nature of these cells, getting access to the DNA is harder than with other samples.

- Cells taken via a buccal swab, which is to say from the inside of the cheeks, is effective (not least because your mouth contains more white blood cells than you probably think). This is the mode of sampling used by 23 & Me and other companies that do not perform full-genome sequencing. The problem is that your mouth is very far from sterile, and there are problems associated with bacteria and other foreign material contaminating the cells.

What we want for purposes of extracting DNA is blood. Although red blood cells don't have nuclei and therefore don't contain any nuclear DNA, white cells do. As well, unless you are very unwell, blood is sterile. It is possible to extract blood from the circulatory system and put it in a sterile vial, keeping it completely isolated from environmental contaminants. That is what you want to get: a sample of blood to send to a genomics laboratory for DNA extraction and sequencing.

People are full of blood, and you need only about 5 mL for genomic analysis. Getting the blood out of you and into a sample tube is not always straightforward, though.

Rules and Regulations

Different countries have different rules about drawing samples of blood. In some places, the process of putting a needle in someone's arm and pulling out a few milliliters of blood is a full-on medical procedure to be carried out only by qualified and licensed medical technicians—they are called *phlebotomists*—acting on the orders of a physician. In other places, it is considered hardly more serious than getting a haircut. Look into phlebotomy (sometimes associated with pathology) in your area to find out what the rules are.

In the United States, as is so often the case, you can do pretty much anything you want if you have the money to pay for it. There are private medical clinics that will draw blood samples, provide all the necessary equipment and qualified people for a reasonable fee, and send you on your way with a sample to use in any way you like. (Literally, any way you like: in researching this book, we found that there is a whole subculture of people with sexual fetishes around blood. They hire phlebotomists for parties. No kidding. It's totally legal in at least some states. You learn something every day.)

In other countries, such as Australia, rules around taking blood are much more restrictive. We live in Australia, and whenever we want a blood sample, we have to go to a doctor and get an order, which we then take to a pathology collection place. The doctor's order has to specify that the samples be given to the patient, rather than to a laboratory. There is a little bit of extra hassle and expense—you have to pay both the doctor and the phlebotomist—but it is all perfectly legal. We found a doctor who was happy to issue the pathology order we needed in exchange for access to the genome information we got back from the genomics lab. You might need to ask around, but plenty of doctors—particularly younger ones—have liberal policies about this kind of thing.

Do-It-Yourself Phlebotomy

You can draw your own blood. The most correct way to do this is to first become a qualified phlebotomist yourself, a process that takes a few dozen hours of training in a trade school, community college, or private training company. There are various certifying authorities, each with their own syllabi and rules about practical experience. (The two most popular ones in the United States are the Certified Phlebotomy Technician [CPT] certificate of the National Healthcareer Association and the Phlebotomy Technician certificate of the American Society for Clinical Pathology). Most courses cover first aid and CPR, as well as venipuncture, and will make sure you are completely *au fait* with hygienic collection practices and the rules and procedures associated with safely handling needles and biohazardous material. If you are the kind of person who likes to take classes and collect certifications, a modest investment of time and money will get you a phlebotomist's certification and a new line item to put on your résumé.

That said, it is not necessary to gain a formal qualification. We live in the golden age of self-education, and there are plenty of excellent tutorial videos on YouTube and elsewhere. Search for *venipuncture tutorial* and *phlebotomy*, as well as *How to draw your own blood* to get started.

The general idea is that you apply a tourniquet to the upper arm, which causes veins to engorge. You sterilize the intercubital fossa—the inside of the elbow—with alcohol or another topical antiseptic and then locate a vein and insert a cannula (a hollow needle) into it. You can then remove the tourniquet and draw blood out through the cannula, into a sterile sample tube. As long as you are careful about sterile technique, the risks are minimal.

You can order the equipment you need to draw blood from eBay, AliExpress, and other online sources. You need these things:

- A *butterfly blood collection set*, which is a cannula connected to a collection fitting (sort of a hollow, sharpened prong) via a length of flexible tubing—all sterile and single-use, of course. The needle has a butterfly-shaped plastic casting a short distance from its point, which enables the person drawing blood to more precisely control the needle and perhaps secure the needle with tape. These cost only about a dollar each, but you typically have to buy a box of 20 or more. Strictly speaking, butterfly-style cannulae are not the only option: you can go with a straight vacuum cannula or a syringe if you prefer, but butterflies seem to be the easiest and most comfortable.

▪ A *vacutainer* or *vacuette*, which is a sort of sterile polyethylene test tube with an airtight cap. The tube has a partial vacuum inside and a self-sealing, puncturable membrane on top. The idea is that you press the vacuette onto a prong at the end of the tube connected to the butterfly needle, which causes the blood to be sucked up into the vacuette. Vacuettes for whole-blood purposes usually have a lavender or pink cap and contain ethylenediaminetetraacetic acid (EDTA), an anti-coagulant and preservative that keeps the blood from clotting or degrading for a period of days. It's hard to buy just one or two of these containers. You usually have to buy a box of 100, but such a box will cost only USD $20 or so. You might be able to re-sell your extras.

▪ A tourniquet.

▪ Topical antiseptic, such as rubbing alcohol or iodine. (Betadine is a brand name.)

▪ Miscellaneous tape and bandages.

It should be noted that while the people on YouTube make drawing blood look easy, it is not always so. We have all experienced, or at least heard about, a nightmare blood-draw session in which the technician drawing the blood cannot locate a vein, the vein moves around, or there is not enough blood pressure to make the vein accessible to the cannula, and the patient ends up with multiple punctures and a painful bruise or two. If that can happen with trained, experienced phlebotomists, it can certainly happen with the home gamer. Beware.

Legal Considerations

We are not lawyers, but as far as we can tell, crimes related to the illegal or unlicensed practice of medicine are concerned with representing yourself *to others* as a doctor or other medical professional and with attempting to diagnose and treat conditions *in other people*. These laws generally do not cover things you do to yourself and probably do not apply if you perform (reasonably simple) procedures on others as long as everyone is aware that you are not actually a trained doctor, nurse, or technician. Drawing blood is not particularly hard or dangerous—though, to be honest, we have not done it—and you will almost certainly be legally safe if you give DIY phlebotomy a try. Use your judgment, and remember, we are not legal authorities in any jurisdiction.

Shipping the Sample

Once you have secured your sample in its vacuette, give it a gentle shake to ensure that the preservative is mixed in with the blood. Then, put it in a refrigerator to cool it off. (Do *not* freeze it.) The idea is that the sample will last longer if it's kept cool.

We had to ship our samples to the laboratory in Hong Kong, a trip of about 36 hours from Sydney via Manila by one of the usual "overnight" courier services. We wanted our sample to arrive as cool as possible, but we didn't want to use dry ice (which complicates shipping) or messy water ice. We had to think of an alternative solution.

In our case, we purchased a vacuum-insulated metal drink bottle with a 500 mL capacity and a wide mouth (Thermos is one brand name, but there are plenty of good-quality makers). We cooled the bottle overnight in our household freezer. Meanwhile, we bought a small quantity of what are called *glass stones* or *vase fillers*, which are flattened glass beads intended to secure the bases of dried and artificial flowers in an arrangement. Don't get confused with "florist's beads," which are soft polymer balls that will retain water and can supply an arrangement of cut flowers with moisture for an extended time—you don't want those. You could use any kind of glass beads or maybe dense polyurethane beads. We put the glass stones in the freezer to cool down as well.

The idea was based on the ideas of *thermal mass* and *thermal inertia*. The concept is that some materials are slow to change temperature (they have a high *specific heat*), and objects surrounded by those materials could be kept at a certain temperature for a relatively long time. Think of a piece of aluminum foil, heated in a flame. Aluminum foil has a low specific heat. It's hot when removed from the flame, but quickly cools down when exposed to room-temperature air. In contrast, a rock (with relatively high specific heat) heated at the bottom of a campfire and removed from the coals will retain its elevated temperature for some time in the ambient air.

We took the cooled vacuum bottle and filled it about halfway with some of the chilled glass stones, then nestled the vacuette, still in its bag, into them vertically. We filled the remainder of the insulated bottle with glass stones, sealed the top, and put the whole thing into the refrigerator to stay cool while we got ready to go.

For the trip to the airport (a drive of about an hour in Sydney's summer heat), we packed the vacuum bottle with the sample and stones inside into an ordinary picnic cooler filled with cold packs. This was probably overkill. Whether the exterior of the vacuum bottle was exposed to ambient air an hour earlier or later wasn't going to make a huge amount of difference, but we wanted to give it the best possible chance of arriving in Hong Kong in top condition.

At the courier's facility near the airport, we filled out the documents specifying the sample's destination, making sure to declare that the package contained a medical sample. We had to specifically declare that the sample was not infectious in a letter that said:

"This is to certify that the shipment contains one sample of 15 mL human blood that is for Research Purpose Only. The sample is derived from a human source, and prepared in a laboratory. All the material is inactivated, and it is not hazardous, not infectious, not HIV positive, not dangerous, not toxic, and not radioactive. The material will be destroyed by autoclave after experiments."

"I hereby agree to send the above materials to BGI Tech Solutions (Hong Kong) Co., Ltd. for research purposes."

We packed the vacuum bottle into one of the courier company's cardboard boxes, which was marked with biohazard stickers. We paid the courier's fee— just over AUD $100—and kept a copy of the shipping document, which included a tracking number. Our sample was on its way.

Receiving the Results

About a month later, we received at our home a portable USB hard drive via the same courier we'd used to ship the blood to Hong Kong. On it, we found just over 400 GB of data.

The hard drive contained three main folders.

- `clean_data`
- `result_alignment`
- `result_variation`

Sequences and Quality Control Information

The `clean_data` folder contained 65 files, being 16 sets of 4 files, as shown in Figure 3.1.

190503_I77_V300018093_L3_B5GHUMihkRAAAAAAA-565_1.fq.gz
190503_I77_V300018093_L3_B5GHUMihkRAAAAAAA-565_2.fq.gz
190503_I77_V300018093_L3_B5GHUMihkRAAAAAAA-565.base.png
190503_I77_V300018093_L3_B5GHUMihkRAAAAAAA-565.qual.png

Figure 3.1: The four raw data files resulting from sequencing cellular material

There also was an index file in Microsoft Excel format. Those sets of data represent the bulk of what we wanted done. The first two are FASTQ (.fq) files that contain long genome sequences. They are very large, many gigabytes in size, and contain the actual genetic sequences that were derived from the DNA in the blood samples we sent. We discuss FASTQ files, their format, and the data they contain in Chapter 9.

The other two files are much smaller. They are PNG images and describe the quality of the sequence information contained in the FASTQ files. These quality control (QC) graphs show the quality of the sequences and indicate how reliable their information is.

Alignment Information

In the result_alignment folder, we found the following five files, plus an Excel index file:

- A BAM (.bam) file, which contains information on alignments. (See Chapter 9 for more information on alignments and the BAM format.)

- A BAM index (.bam.bai) file, which acts as a sort of table of contents for the BAM file. It can be used by genome viewer software such as the Broad Institute's Integrative Genomics Viewer (IGV). With an index file, you don't have to upload a whole BAM file to examine parts of it; the index enables the software to upload only the parts it needs.

- Three graph files that show that the sequencing process achieved a sequencing depth of 35 times in about 95 percent of base pairs, which is almost, but not quite, in line with what we engaged the lab to do. (We asked for 45× depth.) Figure 3.2 shows this information.

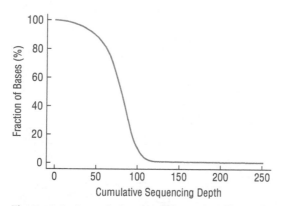

Figure 3.2: A graph showing how many times each base pair in a sequence was examined. The graph shows a drop-off in the percentage of base pairs sampled more than 50 times.

Variation Information

The `result_variation` folder contained the following four subfolders:

- The folder `cnv` contains information about copy number variation, which is duplication or deletion of a large chromosomal region.

- The folder `indel` contains information on inserts and deletions contained in VCFs. Inserts are (usually relatively short) multinucleotide sequences put into the genome in the middle of the "normal" sequence; deletions are chopped-out sections.

- The folder `snp` holds VCF files describing SNPs.

- The folder `sv` contains information on structural variations, which are large (typically over 1 kbases) variations affecting the genome, such as inversions (in which a sequence is reversed) or translocations (in which a portion of one chromosome attaches to another one).

The CNVs and SV information should be taken with a grain of salt, as whole-genome sequencing is not very good at identifying them and most of the software calls are false positives.

Summary

The purpose of this chapter was to show you how to take a blood sample and use a service laboratory to get FASTQ, BAM, and VCF files you can use for further analysis. Now that we have crossed the line from organic tissue to pure data, we can begin working with the genomic information using computational resources.

The Bioinformatics Workflow

This chapter is concerned with the overall workflow of bioinformatics, in which we go from a tissue sample all the way to a meaningful genome as complete as possible, identifying the places where the test subject's genome differs from the rest of the species.

We cover the process of taking some biological material—some cells—and extracting their genetic material. We explain how to convert the extracted genetic material—the DNA, in the case of most cells from complex organisms—and convert that into data that can be manipulated by an electronic computer. Following that, we show an overview of the bioinformatics workflow that will enable us to derive meaningful information from the data.

The process begins in a "wet lab," with manipulation of actual cells. (We'll focus on human blood.) From there, a sequencing machine will convert the cells to data—an identified sequence of nucleotides, complete with a statement of how likely each nucleotide identification is to be accurate. After that, it's a matter of cleaning up the sequence to remove as many of its flaws as possible, before aligning it to a reference genome. With the alignment complete, it should be possible to see how the person from whom the sample was taken differs from the species average—and that variation is where the value of genomics is.

In this chapter, we will look at the essentials of practical genetics and bioinformatics. We will examine the processes, tools, and forms of data that enable us to carry out genetics work.

Extraction of DNA

We saw in Chapter 2 that DNA is a molecule—the famous double helix—that exists in the nucleus of cells. There, the DNA is arranged into chromosomes, where it is tightly spooled around proteins called *histones*.

Deriving Nucleated Cells from Whole Blood

To get DNA, we must first begin with a sample of nucleated cells—that is, cells that have nuclei. Most cells do have nuclei, so it is possible to extract DNA from almost any tissue sample. As an aside, in detective novels and cop shows in which forensics experts are depicted deriving DNA from such materials as urine or the seminal fluid of a man who has had a vasectomy, the creators are not totally off base. Bodily fluids such as those typically contain whole cells that slough off from inside the body, and DNA can be extracted from those cells. Saliva contains a certain number of cells that drop off from inside the mouth, which is why a lot of mass-market genetics services require saliva samples.

For the purposes of this chapter, we will assume that we are working with a blood sample from a human. Interestingly, red blood cells do *not* have nuclei—they've evolved that way so as to fit more oxygen-carrying hemoglobin within the double-layer lipid membranes that form their cell walls. Similarly, platelets—essential to stopping bleeding by coagulation—do not have nuclei either.

There is more to blood than red cells and platelets, however. White blood cells, which are part of the body's immune system and are also known as leukocytes, have nuclei and are where DNA comes from when starting with a sample of blood.

A sample of whole blood, broken down by volume and disregarding normal variations between men and women, is typically 40–50 percent red blood cells, 1 or 2 percent white blood cells and platelets, and the rest plasma (which is water with various hormones, clotting factors, and other trace substances present in it).

So the first thing to be done in isolating DNA from whole blood is to separate the white blood cells from the rest of the blood. This is typically done with a centrifuge, in which a spinning force causes cells to settle out of the plasma. After fractionation by centrifuge, a formerly uniform sample is layered, with a thick stratum of red blood cells at the bottom, a thick layer of yellow plasma at the top, and a thin layer comprising white cells and platelets between the two thick layers.

Called the *buffy coat* because of its off-white color, the middle layer is extracted from the rest of the fractionated blood sample with a pipette and used as the basis for the rest of the DNA extraction process.

Processing Nucleated Cells

The next step in processing a sample of biological material into data that can be manipulated by a computer is to process the cell sample into DNA, as pure as possible. Simply put, this is a process of breaking open the cells and their nuclei and then washing away all the miscellaneous cellular material, leaving only DNA.

You can think of cells as waxy bags filled mostly with fluid. The cell membrane—the outer boundary of the cell—is formed by a double layer of molecules called *phospholipids*, each of which has a hydrophilic head and a hydrophobic tail. They arrange themselves tail to tail, with the water-loving heads pointed either out into the blood plasma or inward toward the cytoplasm that fills the cell.

Within the cell membrane, in the cytoplasm, various structures exist: the mitochondria, the ribosomes, the Golgi apparatus, and so on. Also in the cytoplasm is the nucleus, which is another waxy balloon, but with two walls (each of which has two layers of phospholipids arranged tail to tail). The chromosomes, made up of histones and tightly coiled DNA, sit within the nucleus.

To get at the DNA, we must break down the cells physically—tear open the cellular bags to release the genetic material in a process called *lysis*. There are various ways to do this, but it's most often done by physical means—simply putting the cells in what is basically a blender or grinding them with a mortar and pestle. Lysis can also involve detergents. (After all, the goal is to break down the fatty cell walls, not unlike what happens when you remove grease from dishes and utensils.) An enzyme called *Proteinase K* can also help melt down cellular structures.

The next phase in purifying DNA is to take the cellular mush and process it further. First, positively charged sodium ions are added to the material and mixed well. The positive ions attach themselves to the DNA molecules, which carry a negative charge. Neutralizing the DNA makes it more chemically stable and less soluble in water.

A cold alcohol (usually ethanol or isopropanol) is then added to the cellular material. With mechanical manipulation (i.e., stirring), the DNA will precipitate out of the rest of the mush—it can be visible as stringy, white material—and can be removed from the rest of the matter with a pipette. Further rinsing with alcohol may be needed to wash away other cellular debris.

Finally, the clean DNA is put into sterile, slightly alkaline water for stable short-term storage and further examination.

FASTA Files

The output of a sequencing operation, in its simplest form, is a FASTA file. A FASTA file—the name derives from an old piece of software and doesn't really stand for anything—is a file that contains a sequence of nucleotides (A, C, G, and T). In truth, the FASTA format has applications beyond nucleotides and can be used to encode information about amino acids and proteins, but we'll focus here on its applications in recording genome sequences.

A FASTA file is a text file encoded as ASCII text, which makes it easy to manipulate with UNIX shell tools and other text-manipulation software (such as Perl, Python, and anything else that implements regular expressions).

The FASTA format is straightforward. For each sequence, there is a header line that begins with a > character. After that, the sequence itself appears as a series of ASCII characters (just the four of them in the case of nucleotide sequences). Here's a sample from a human reference genome:

```
> chr1
GGTGTAGTGGCAGCACGCCCACCTGCTGGCAGCTGGGGACACTGCCGGGCCCTCTTGCTCCAACAGTACTG
GCGGATTAT
AGGGAAACACCCGGAGCATATGCTGTTTGGTCTCAGTAGACTCCTAAATATGGGATTCCTGGGTTTAAAA
GTAAAAAATA
AATATGTTTAATTTGTGAACTGATTACCATCAGAATTGTACTGTTCTGTATCCCACCAGCAATGTCTAGGA
ATGCCTGTT
TCTCCACAAAGTGTTTACTTTTGGATTTTTGCCAGTCTAACAGGTGAAGCCCTGGAGATTCTTATTAGTGA
TTTGGGCTG
GGGCCTGGCCATGTGTATTTTTTTAAATTTCCACTGATGATTTTGCTGCATGGCCGGTGT
```

(In some FASTA files, the sequence is broken into 80-character rows, but that's not required by the format specification and is not always the case.)

The problem with FASTA files is that, in most cases, every nucleotide in every sequence is not 100 percent authoritative (reference genome sequences being a notable exception). For that reason, we need a way to attach a description of quality to each nucleotide. That's the purpose of the next file format we cover in this chapter: FASTA files with quality information, also known as *FASTQ files*.

FASTQ Files

Files in the FASTQ format are essentially the raw data that results from a modern genome sequencing operation for anything other than a reference genome. They are the output of a sequencing machine, such as those from Illumina and Oxford Nanopore. A FASTQ file is where "wet lab" processes end and manipulation of genetic information by computers begins.

A FASTQ file contains sequences of letters representing nucleotides (A, C, G, and T), a quality score for each letter, and some labeling information. The quality score is necessary because the sequencing process is imperfect. For a given nucleotide, the sequencing machine is making a "best guess" as to what that nucleotide really is. The guess is considered better if the conditions under which the guess (or "call") was made were closer to ideal, and the call is deemed worse if the conditions were less good.

Quality is the probability that a given read is correct. It is determined by the sequencing machine, which looks at the images of genetic material it used to make a decision about whether a given nucleotide was A, C, G, or T. It considers the resolution of the image and also its sharpness. A high-resolution image with strong sharpness could be relied upon better to give a correct read, so a nucleotide determined with such an image would be assigned a high Phred score. A low-resolution image, or one with poor sharpness, could not be relied on as much, so a nucleotide associated with an image like that would get a low Phred score. The actual algorithms used to evaluate image quality and assign a corresponding quality score are proprietary to the makers of sequencing machines.

FASTQ files are ASCII-encoded text files, which means it is easy to manipulate them using regular expressions, standard command-line tools, and languages such as Perl, Python, Ruby, and R. The format evolved from an older standard called *FASTA*, discussed earlier in this chapter, which allowed sequences to be split across multiple lines to accommodate the limitations of older computing equipment. FASTQ files, in contrast, have exactly four lines per sequence. A typical set of four looks like this:

```
@cluster_36:UMI_AACAGA
TCCCCCCCCCAAATCGGAAAAACACACCCCC
+
5?:5;<02:@977=:<0=9>@5>7>;>*3,-
```

The first line is a label, identifying where the sequence came from. It always begins with an @ character.

Note that the second line and the fourth line contain precisely the same number of characters. There is a reason for that.

The second line is the actual nucleotide sequence read by the sequencing machine. The four letters represent the four types of nucleotide: adenine (A), cytosine (C), guanine (G), and thymine (T).

The third line always begins with a + character and may (according to the FASTQ specification) contain a repeat of the label that appears in the first line. However, this second appearance of the label is usually omitted in the interest of keeping file size down.

The fourth line comprises a series of characters, each of which represents the quality of the corresponding data in the second line. Each character represents the probability that the nucleotide specified in the same position on the second line was correctly determined by the sequencing machine. The details of exactly how the encoding works appear later in this chapter.

Phred Scores

The characters in the fourth line of a FASTQ sequence description represent quality scores. In other words, they represent the probability that the sequencing machine correctly identified the nucleotide (A, C, G, or T) it denoted in a given location.

Ordinarily, such a quality or confidence score would be specified as some sort of probability or percentage: there is a 99.999 percent probability that the nucleotide has been determined correctly, or there is a 90 percent chance that it has been determined correctly. In genomics, however, and particularly in FASTQ files, quality is represented with a Phred score, also known as a Q *score*.

(Why is the Phred score called that? It was originally calculated by a piece of software called Phred, which implemented an algorithm also called Phred. These were developed in the 1990s by a team led by Phil Green, a professor at Washington University in St. Louis—so maybe "Phred" is easier to say than "Phgreen"? That's pure speculation. The truth seems to have been lost to history.)

A Phred score is calculated with the following formula:

$$Q = -10 * \log_{10} p$$

In the formula, Q is the Phred score, and p is the probability that the value is wrong, expressed as a real number between 0 and 1. Let's work through a sample.

To express a 99.99 percent confidence that the data is correct, use 0.001 (the probability that it is incorrect, or 1 − 99.99) as your value for p.

$$Q = -10 * \log_{10}(0.001)$$
$$Q = 30$$

This is fine, except that the value 30 has two characters. In FASTQ, a one-to-one correspondence between a single nucleotide and its quality score is needed. There has to be a way of converting two-place Q scores, such as 30, into a single character.

By the way, in genomics, a Phred score of 20 or greater, which means there is a chance of 1 percent or less that the nucleotide has been named incorrectly, is considered good.

ASCII Encoding of Phred Scores

Recall that the American Standard Code for Information Interchange (ASCII) specification is a way of encoding printable and other characters with bytes of data. It's very limited—it doesn't support accented characters, let alone non-Latin character sets—and has been supplanted in most modern situations by the more capable Unicode standard. ASCII works well in this corner of genomics, though, in that it provides a simple way to encode a two-digit Phred score as a single character.

In ASCII, characters correspond to integers (0 through 127), but not all characters are printable. ASCII character 13 represents the carriage return instruction, for example. The range of printable characters (excluding the space character) is 33 (representing the exclamation point) through 126 (the tilde)—a total of 94 printable characters. That means FASTQ can use those 94 printable characters to encode 94 Phred scores, which is enough for all practical purposes.

To convert a Phred score to an ASCII value, add 33 to it. A Phred score of 2 corresponds to ASCII 35, which can be rendered as the # character—just the one character. The real value of ASCII encoding becomes evident with larger Phred values—say, 20 or greater (which are values indicating a high probability that the nucleotide was named correctly by the sequencing machine). A Phred score of 80 (representing a 99.999999 percent probability of a correct nucleotide) corresponds to ASCII 113, which is the character q.

Consider again the sample set of four FASTQ lines we looked at earlier in this chapter:

```
@cluster_36:UMI_AACAGA
TCCCCCCCCCAAATCGGAAAAACACACCCCC
+
5?:5;<02:@977=:<0=9>@5>7>;>*3,-
```

We can decode the Phred scores now. You will see that the first nucleotide, named T (in bold), has a corresponding Phred score that is encoded as the character 5. That character has the ASCII code 53. With a quick calculation, 53 – 33, we see that the Phred score is 20. If we want, we can convert that to a probability of accuracy by using the Phred score formula.

$$Q = -10 * \log_{10} p$$

If we plug in 20 as Q and solve for p, we get a probability of inaccuracy of 1 percent, which is the same as probability of accuracy of 99 percent.

Similarly, we can see that the quality score of the C in bold in the 28th position is quite low. Its Phred score is represented by the * character (ASCII 42), which corresponds to a Phred score of 9 (42 – 33). With the equation, we can work out that the probability that the C shown represents the truth is only about 87 percent.

Note that the ASCII range of printable, nonspace characters (33 to 126) includes the @ character and the + character. Because the first line in a set of four lines describing a sequence in a FASTQ file begins with @, and the third line in such a set begins with +, you need to be careful using grep and other text tools to look for those characters. You can't count the number of sequences in a FASTQ file by looking for @ characters at the start of lines, because while it is true that the first line in a set will always begin that way, the fourth line can also. This will happen if the Phred score of the first reported nucleotide is 31 (corresponding to 64 − 33, where 64 is the ASCII code of the @ character).

Figure 4.1 displays the Phred score or quality score (Q), the associated probability of an error (P error), and the associated ASCII code (based on ASCII base 33).

Phred	P error	ASCII	Phred	P error	ASCII	Phred	P error	ASCII	Phred	P error	ASCII
0	1.00000	33 !	11	0.07943	44 ,	22	0.00631	55 7	33	0.00050	66 B
1	0.79433	34 "	12	0.06310	45 -	23	0.00501	56 8	34	0.00040	67 C
2	0.63096	35 #	13	0.05012	46 .	24	0.00398	57 9	35	0.00032	68 D
3	0.50119	36 $	14	0.03981	47 /	25	0.00316	58 :	36	0.00025	69 E
4	0.39811	37 %	15	0.03162	48 0	26	0.00251	59 ;	37	0.00020	70 F
5	0.31623	38 &	16	0.02512	49 1	27	0.00200	60 <	38	0.00016	71 G
6	0.25119	39 '	17	0.01995	50 2	28	0.00158	61 =	39	0.00013	72 H
7	0.19953	40 (	18	0.01585	51 3	29	0.00126	62 >	40	0.00010	73 I
8	0.15849	41)	19	0.01259	52 4	30	0.00100	63 ?	41	0.00008	74 J
9	0.12589	42 *	20	0.01000	53 5	31	0.00079	64 @	42	0.00006	75 K
10	0.10000	43 +	21	0.00794	54 6	32	0.00063	65 A			

Figure 4.1: Phred scores

Alignment to a Reference Genome

With the creation of a FASTA or FASTQ file, we have a collection of nucleotide sequences. Sequences from Illumina and BGI typically range from about 50 to 150 nucleotides in length, though nanopore, PacBio, or Chromium 10x sequencing—relatively new techniques—can yield longer continuous sequences. Remember, though, that all of these sequences are in no particular order. They came from fragments of DNA that were taken from cellular material that had been chopped up mechanically and chemically—and even then, any long DNA segments that remained might have been chopped up further because of the limitations of the sequencing technique being employed.

The contents of a FASTA or FASTQ file, therefore, are not useful by themselves. To derive any useful information from the newly sequenced DNA, a researcher has to try to assemble the fragments into something like a complete DNA strand. Ideally, it would be possible to put all the fragments together into 46 long sequences that matched the sequences of the 23 pairs of chromosomes belonging to the sampled individual.

The short reads—millions of them typically—do not contain any position information. It is not possible, just by looking at a fragmentary sequence, to tell anything about where it fits in the genome—not even which chromosome it came from. So, to assemble the fragments into a complete sequence, we need to use a *reference genome*.

The value of a reference genome lies in the fact that, for a given kind of organism, the genetic variation across individuals is very small. In humans, who have genomes about three billion nucleotides long, the variation is about one-tenth of 1 percent—about three million differences among the overall three billion nucleotides. That means that if we have a reference genome, we can fit the fragmentary sequences we've derived from a sequencing process into similar regions of the reference genome. It's sort of like one of those puzzles for little kids that have cut-out spaces for the individual pieces to fit into: the pieces don't interlock with each other but rather fit into a matching void.

The process of fitting sample sequences onto a reference genome is called *alignment* or *mapping*.

On the other hand, except for identical twins, no two individuals have exactly the same genetic sequence (even identical twins will differ genetically, practically speaking, due to each having different mutations and viral infections). Therefore, matches between the sampled genome and the reference genome will not be exact. Indeed, if they did match exactly, there would be no point to genetic sequencing. The whole reason for doing sequencing of an individual is to identify differences between that person's genome and the reference genome.

Matching short samples to the reference genome is hard for several reasons. First among these reasons is that the basic task is to match sequences of a few hundred letters into a reference genome consisting of three billion letters. It's just a monumental task, even for a computer, to try to find where in such a gigantic string of characters a short string might fit.

Beyond that, there might be several places a given sample sequence might match with the reference genome. In particular, there are places in the genome that feature a lot of repetitions. A given sample sequence would fit with any of several similar sequences, and there's no way to tell which.

Furthermore, the whole point of genetic sequencing is that the matches are not going to be perfect—nucleotides will differ between the reference genome and the individual's genome. We are not clones, after all. On the other hand, not all variations are "real." They may instead be an artifact of the sequencing process, which is one of the reasons why we have a Phred quality score—to distinguish real genetic variations from process-induced variations.

The mismatches aren't just limited to differences between the nucleotides at given locations either. It is possible for the sample sequence to have extra nucleotides inserted at a given location (an insertion) or a series of nucleotides deleted (a deletion). These phenomena are collectively referred to as *indels*, and they mean that a given sample sequence and the nominally matching region on the reference genome need not be the same length. The alignment process has to allow for this possibility.

Reference Genomes

What, then, is a reference genome?

A *reference genome* is a "standard" sequence of nucleotides, representing a single species, published and made available for comparison against other individual genomes. Generally, a reference genome represents some sort of compilation of the genomes of several individuals. Several individuals' genomes are combined to prevent a bias from emerging with respect to what characteristics are "normal" within the species. The current human reference genome consists of a mosaic of genomes from more than 10 anonymous individuals, males and females, though the main contribution comes from a man of African-European heritage referred to as RP11.

The genomes of all healthy humans include two copies of each autosomal chromosome (that is, chromosomes 1 through 22) and two sex chromosomes (two X chromosomes in females and one each, X and Y, in males). In contrast, a reference genome includes only one copy of each autosomal chromosome as well as one copy each of X and Y—that's a total of 24 reference chromosome sequences.

As of this writing, the latest standard human reference genome is called Genome Reference Consortium Human Build 38 (GRCh38), which was released in December 2013. Beyond that, there are minor updates to the reference genome, and the most recent of those is Patch 14, released in February 2022—so the full name of the current human reference genome is GRCh38.p14. It contains just over three billion nucleotides (with gaps) and just under three billion without gaps. More than half of the genome is in continuous sequences (*contigs*) greater than 67 million nucleotides in length. It includes representations of 24 chromosomes, being the 22 autosomes and the two different sex chromosomes, X and Y.

As well, GRCh38 includes sequences of mitochondrial DNA. Recall from earlier in this book that mitochondrial DNA (mtDNA) is passed exclusively from mother to child and is therefore an essential tool in the study of population genomics—and by extension the migration of populations around the world, which is relevant to anthropology.

Prior to GRCh38 was GRCh37, which the Genome Reference Consortium (GRC) released in February 2009. Some databases still refer to GRCh37, but most have migrated, or are in the process of migrating, to GRCh38. The newer reference genome has a better coverage of the human genome than the older reference GRCH37 and in particular now incorporates sequences of the centromere (central region of the chromosome attached to fibers during mitosis). It also includes many additional sequences (called *alternate contigs*, or *alt-contigs*) corresponding to areas of high variability among individuals, in particular areas related to the immune system, and has fewer gaps than the previous reference (about 500 gaps versus more than 150,000 gaps in the previous reference). Globally, evaluations of the two builds showed that a greater proportion of sequenced reads mapped to the new GRCh38 build than to the older GRCh37 build.

Reference genomes come in the form of a series of FASTA files (not FASTQ files, of course, because a reference genome is considered authoritative and there's no need for quality scores). You can download reference genome FASTA files several places, including the following:

- The National Center for Biotechnology Information (NCBI), which is part of the U.S. National Institutes of Health (NIH)

- The University of California at Santa Cruz (UCSC)

- The Ensembl genome database project, which is a collaboration between the European Bioinformatics Institute and the Wellcome Trust Sanger Institute

Quality Control

Once we have a FASTQ file, we can examine it to see which parts of it are of sufficient quality to potentially yield meaningful results. In examining a sequence, we must consider the following:

- **Nucleotide quality**: We must look at the Phred scores to see which of the stated nucleotides has sufficient probability of correctness to be useful. In general, we disregard reads (reported nucleotide identifications) that have a Phred score below a certain value (often 20, denoting a probability of error greater than 1 percent and denoted by ASCII encoded score 5).

- **Nucleotide distribution**: The frequency of each kind of nucleotide should be approximately equal across the entirety of the sequence. That is, there should be approximately equal numbers of A, C, G, and T reads throughout the stated sequence. Certain sequencing techniques can lead to biases, though—particularly at the beginnings and ends of sequences. One function of quality control is to account for those biases. The presence of too many G and C calls (high GC content), in particular, can indicate sample contamination or other problems.

- **Sequence duplication levels**: They can arise from biased PCR enrichment, which is a concern, or from genuine over-representation of some sequences (relatively rare in DNA sequencing, though it can happen that, by chance, sequences are cut at exactly the same positions; it's more common when doing RNA sequencing, where some transcripts can be highly expressed).

The standard tool for looking at a FASTQ file and reporting on its quality is called *FASTQC*, an open-source package freely available from the Babraham Bioinformatics site.

```
www.bioinformatics.babraham.ac.uk/projects/fastqc
```

The output of FASTQ is an HTML file that can be viewed in your web browser.

What can you do if your genome data is found to have quality problems? One common and useful strategy is to excise low-quality regions from the sequences. This process is called *trimming*.

Trimming

Trimming a genome sequence is the process of cutting out areas that have been determined to be of low quality for whatever reason. Trimming has to be done in moderation, because cutting out too much will lead to problems later when there is insufficient data to yield meaningful analysis, while cutting out too little will likewise lead to trouble later because the data is of too low quality.

Trimming should be used to eliminate *adapters*, which are occasionally present in the FASTQ file. Adapters are strings of synthetic nucleotides of known sequence. They contain information useful for the sequencing process: barcoding sequences, binding sites for the sequencing primers, and binding sequences that immobilize fragments in the flowcell of the sequencing machine. After DNA fragmentation, adapters are ligated to each DNA fragment. Adapters are not usually present in the FASTQ file, as the sequencing starts after the 5′ primer. But if some DNA fragments to sequence are too short, the sequencing can continue till the 3′ primer and into the 3′ adapter. The sequence of the 3′ adapter then contaminates the FASTQ file and should be trimmed out.

Trimming is usually done with a piece of Java software called Trimmomatic. You can download Trimmomatic from the Usadel Lab site.

```
www.usadellab.org/cms/?page=trimmomatic
```

With Trimmomatic in place, you can clean up your sequences. A typical trimming work sequence is as follows:

1. Remove adapter sequences.
2. Remove leading low-quality bases—say, below Phred code 5 (that is, Q score 20).
3. Remove leading N bases (N being FASTA/FASTQ code for "any nucleotide"—either A, C, G, or T—and therefore valueless).
4. Remove trailing low-quality bases.
5. Remove trailing N bases.

6. Scan the read using a sliding window four bases wide, and trim when the average quality for the four is less than 15 (Phred 0).

7. After all that cutting is done, drop any sequence that is not at least 36 bases long.

At the end of trimming, the idea is that you have fewer nucleotides in the sequences that remain, but they are of higher quality and will yield more meaningful information later in the process.

The Alignment Process

The *alignment process* is the process of matching fragmentary sequences from a sample (a test subject) with a reference genome. You take your sample sequences—having been encoded in a FASTQ file and trimmed to improve quality—and attempt to fit them to match a reference genome.

You perform alignment with a piece of software called, logically enough, an *aligner*. The two most popular aligners are BWA (an implementation of three different variations of the Burroughs-Wheeler algorithm) and STAR (which stands for "spliced transcripts alignment to a reference"). BWA is used for aligning DNA sequences, while STAR is preferentially selected for aligning spliced mature RNA. Alignment is compute-intensive and is perhaps the part of computational genome analysis that is best suited to Amazon Web Services, with its virtually unlimited computing power.

We prefer BWA and so will focus on that package here. BWA implements three algorithms.

■ BWA-backtrack is best suited for short sequences of up to 100 base pairs.

■ BWA-SW can handle longer sequences (up to 1,000 base pairs), but it is slower than BWA-MEM and yields less accurate results. On the other hand, it may be better at handling sequences with many gaps.

■ BWA-MEM is the best of the BWA algorithms, capable of handling both long (up to 1,000 base pairs) and shortish (70–100 base pairs) sequences efficiently and accurately.

BWA takes FASTQ files as input files and generates as output a sequence alignment and map (SAM) file. As the SAM file is very large, it is then compressed into a binary alignment and map (BAM) file.

The most frequently used DNA sequencing platforms involve paired reads. This involves a fragment of DNA (e.g., 750 nucleotides long) that is sequenced a short way from either end (e.g., 150 nucleotides from either end). This results in two different sequencing reads—called *paired reads*. They are related because they are on either end of the same DNA fragment. This information about the pairing of reads is recorded in BAM files and used in analysis. For instance, normally the read pairs map to the reference genome at a certain distance from each other on the same chromosome (e.g., 400 nucleotides distance between them). If the mapping distance is much longer, then there may be a deletion in the sequenced genome. If the mapping distance is much shorter, then there may be an insertion in the sequenced genome. If the sequenced genome is a tumor and one of the read pairs maps to one chromosome while the other read pair maps to another chromosome, then this may represent a translocation where a piece of one chromosome has been aberrantly inserted into another chromosome. Paired sequencing reads have the same sequence identifiers for the two reads. If paired read sequencing technology was used, then there will be two FASTQ files. The second FASTQ file will contain the sequencing reads in the same order, with the same sequencing read identifier, as the first FASTQ file. By convention, the filenames will have the same prefix with R1 and R2 to distinguish them (e.g., `my_fastq.R1.fastq.gz` and `my_fastq.R2.fastq.gz`). Figure 4.2 displays the alignment of the paired reads to the reference genome.

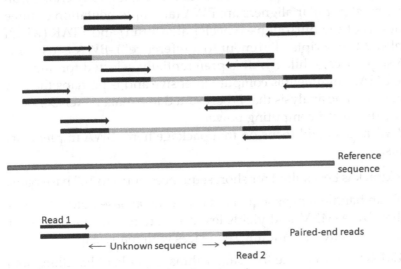

Figure 4.2: Alignment of paired reads to the reference genome

The output BAM file generated by BWA adds alignment information to each sequence of the FASTQ files. The alignment information consists of a tab-delimited line with the following fields:

- Sequence_identifier (identical for the two paired reads, so this read and its mate)
- SAM flag
- Chromosome that this read has been mapped to

- Position in the chromosome that this read has been mapped to
- Mapping quality
- CIGAR string—number of nucleotides that match the reference sequence (M), correspond to insertions (I), deletions (D), or do not match the reference sequence (S).
- Chromosome that the read mate of this read's read pair was mapped to—this and the next two fields are how the BAM file records how far apart are the mappings of the two read sequences of the same sequenced DNA fragment; this field contains the equal sign (=) when it is mapped to the same chromosome
- Position in the chromosome that the read mate of this read's read pair was mapped to
- Length of the DNA fragment that this read pair must have come from, from the beginning of the first read to the end of the second read

The definition of BAM file fields is found on page 6 of the BAM file specification at `samtools.github.io/hts-specs/SAMv1.pdf`. There may be other fields on the same line after the previously listed fields. The SAM flag field values are explained in the BAM file specification, but it is easier to go to `broadinstitute` `.github.io/picard/explain-flags.html`, enter the SAM flag value, and let that website explain that particular value.

Marking Duplicates

As described previously in the FASTQC sections, PCR duplicates are sequence reads obtained by sequencing several copies of the same DNA fragment.
The following are the two most popular tools for marking duplicates:

- Picard MarkDuplicates (`broadinstitute.github.io/picard/`)
- Biobambam (`github.com/gt1/biobambam2/releases`)

Biobambam supports multithreading and is therefore faster than Picard.

Recalibrating Base Quality Score

This data preprocessing step is optional. It consists of looking for systematic errors made by the sequencer when it assigns a quality Phred score to each nucleotide call. Base quality score recalibration (BQSR) applies a machine learning to model these errors and slightly readjusts the quality scores.

Calling SNVs and Indel Variants

Our BAM file is now ready to call the variants: single nucleotide variants (SNVs), small insertions and deletions (indels), structural variants (SVs), and copy number variants (CNVs).

We have our hundreds of thousands of sequencing reads aligned to the human reference genome in the BAM file, like jigsaw pieces placed in the correct place in a jigsaw puzzle. When sequencing depth is 30x, then for each nucleotide position in the genome, there are around 30 sequencing reads that cover that nucleotide position. So, to continue the jigsaw puzzle analogy, it is as if each position in the puzzle has 30 identical pieces on top of it. If a sample actually has a difference at that position on one of the two copies of the chromosome, then half of the sequencing reads will have a different value at that position, and a heterozygous SNV should be reported. If a sample actually has a difference at that position on both copies of the chromosome, then all of the reads will have the different value at that position, and a homozygous SNV should be reported. If a sample has, say, three nucleotides missing at a given position on one copy of the chromosome, then half of the sequencing reads will contain the text "3D" (signifying a three-base-pair deletion) in the BAM file's cigar string at that position, and a heterozygous small deletion indel should be called. If a sample has a one-nucleotide insertion at a given position, then half of the sequencing reads will contain the text "1I" (signifying a one-base-pair insertion) in the BAM file's cigar string at that position, and a heterozygous small insertion indel should be called.

There are various software modules and pipelines that have been developed for processing DNA sequencing reads.

The SNVs and indels are most commonly called by the Genome Analysis Toolkit (GATK) software developed by the Broad Institute, a leading genomics research institute (`github.com/broadinstitute/gatk/releases`).

We will broadly follow the best practices developed and recommended by Broad Institute. You can search for "Broad Institute best practices" and find more details about Broad best practices at `gatk.broadinstitute.org/hc/en-us/sections/360007226651-Best-Practices-Workflows` and videos at `www.broadinstitute.org/partnerships/education/broade/best-practices-variant-calling-gatk-1`.

Sometimes it is not clear-cut whether a variant is present or whether the differences seen are simply errors. Sequencing can produce systematic errors in addition to random errors. The GATK best practices pipeline contains extra steps producing extra information so that the data of multiple samples can be compared at once. This allows the software to do a better job of detecting whether differences are real variants or are only errors. When the variants of multiple samples are called in this way, it is referred to as *joint-calling*. In our book, we have only one sample running through the multistep best practices pipeline. However, if you were joint-calling multiple samples, you would use the same multistep pipeline presented here.

GATK produces a Variant Call Format (VCF) file containing the SNVs and indels. The definition of the fields of the VCF file is found at `samtools.github .io/hts-specs/VCFv4.2.pdf`. For each variant, the chromosome and position of the variant appear in the CHROM and POS fields, respectively. If this variant has already been recorded in the dbsnp database, then its dbsnp identifier appears in the ID field. The value of the nucleotide in the reference genome appears in the REF field. The value of the nucleotide observed in the BAM file appears in the ALT field. The 10th field contains colon-delimited data, and the first piece of data is the genotype of this variant. If the variant is heterozygous (present on only one of the chromosomes) in this sample, then that data is "0/1." If the variant is homozygous (present on both chromosomes), then that data is "1/1."

As a second step, GATK can recalibrate the variant quality scores during a process called *variant quality score recalibration* (VQSR).

Annotating SNVs and Indel Variants

We now have our sample's variants—SNVs and indels—recorded in the VCF file. Each human has around six million SNVs and indels that are different from the reference genome. Most of them are not problematic. However, a few of them might be associated with a problem. How can we tell which variants in the VCF file might be associated with problems? We need to annotate the variants and then look at variants that have annotations that indicate a possible problem. Annotations can, for example, report whether a variant is in or near a gene or is inside the protein coding region of a gene. Annotations can include the consequence of a variant. For example, if a variant changes the amino acid, then it is a called a *missense variant*. If a variant changes the nucleotides that tell the ribosome to stop synthesizing the protein such that it no longer stops, then the variant is called a *stop_lost variant*. Other types of annotations are available. For instance, it is of interest to know whether the variant is rare or common, because a rare variant might be associated with a rare disease, whereas a common variant is almost certainly not associated with any rare and serious disease. The ExAC resource has sequenced the DNA of around 60,000 people and has counted how many of those people have the variants seen in those people. If a variant is not present in ExAC, then the variant is so rare that in a collection of 60,000 people it was seen not even once. Annotation of our variants with ExAC frequencies is available and interesting to do so that for each of our variants we can see whether the variant is common or rare. ClinVar is a database where genetics clinicians and researchers have deposited interesting variants that they have seen in their patients. A variant in ClinVar is considered interesting because either it is associated with a disease or it was thought to possibly be associated with a disease but upon further investigation was found to not be associated with the disease after all. Variants in ClinVar are classified with a clinical significance, such as "Likely pathogenic," "Likely benign," or "Uncertain significance." Annotation of our variants with ClinVar is available and of great interest because it is important to know whether one of our variants is present in ClinVar with a clinical significance of "Pathogenic." Other interesting annotations include pathogenicity scores, such as CADD. CADD scores provide an estimated pathogenicity score for every nucleotide in the human genome, estimated using machine learning techniques. REVEL scores are another set of pathogenicity scores, derived from the results of multiple other pathogenicity prediction scores.

A popular, well-documented, and frequently updated software for annotating human variants in a VCF file is VEP. The base VEP installation provides some annotations, whereas others are provided by VEP plugins. Installation instructions for VEP are available at `m.ensembl.org/info/docs/tools/vep/script/vep_download.html`.

Prioritizing Variants

Each individual has around six million variants, and we can't look manually at them all. If we are looking for possibly harmful variants, then we need a way to prioritize those millions of variants so that we can look at only the most consequential variants. Each bioinformatics team has its own methods, and in many cases, its own in-house software, for automatically prioritizing variants and extracting only the variants that might be harmful. We use one of the publicly available solutions that can be quickly and easily installed and run: the VPOT software (`github.com/VCCRI/VPOT/`).

Inheritance Analysis

When a child has a genetic problem and the parents don't have the problem, sequencing all three provides greater analysis power because we can look for variants that are in the child and are not in the parents. There are a few inheritance models to consider that would result in the child having a genetic disease while the biological parents do not. By sequencing and comparing the variants of the child to those of the parents, we have a much greater chance of identifying the pathogenic variant in the child than if we sequence only the child. There are around six million variants in the child. However, there are only around 20,000 variants that are in the child and are not in the parents, significantly cutting down the number of the variants to evaluate and score for pathogenicity. If one of the parents has the disease while the other parent doesn't, then sequencing all three allows us to identify variants that are in the child and the affected parent and are not in the unaffected parent. Again, this significantly cuts down the number of variants to consider.

Bioinformatics teams have different methods for carrying out trio analysis. The VPOT software is an easy-to-use and easy-to-run software that can carry out inheritance modeling of trio analysis on a VCF file containing family members. The VPOT software can analyze several inheritance models, including autosomal dominant, autosomal recessive, compound heterozygous, and de-novo and case control scenarios.

Identifying SVs and CNVs

Structural variants are larger variants than SNVs and indels. They are usually defined as variants that are more than 50 base pairs in length and can be hundreds or thousands of nucleotides in length. As you can imagine, SVs have the potential to be harmful because they involve large chunks of DNA. Nonetheless, each person is thought to have more than 7,000 SVs. One way of identifying structural variants is by identifying breakends (BND records in the VCF specification) where DNA sequence in one region is connected to a DNA sequence in another non-adjacent region, indicating a deletion, duplication, inversion, or translocation. Breakend records can also identify inserted DNA sequences. The breakend strategy can identify structural variants down to the exact nucleotide. There are several different programs available that identify structural variants by the breakend strategy. A popular one is Manta.

Another strategy for identifying larger variants is the copy number variant method, which identifies either copy number loss (deletion) or copy number gain (duplication) by identifying regions that have a steady change in depth of sequencing. If the sample was sequenced to a depth of 30x and there is a region where the depth is a steady 15x or the depth is 0 (no sequencing reads in this region), then this should be identified as a copy number loss variant. It signifies that there is a deletion on one or both chromosome copies, respectively. If the sample was sequenced to a depth of 30x and there is a region where the depth is a steady 45x, then this should be identified as a copy number gain variant. It may be due to a tandem duplication of this region, or this region has been copied and inserted elsewhere. CNVs don't identify the exact nucleotides where the variant starts and ends. However, this is usually not necessary for interpreting the effects of the variant, given that the variant is large, making it obvious when a gene is affected by the large variant. There are many programs available to identify CNVs, and a popular one is Cnvnator.

For very large copy number variants, traditionally these have been identified by microarray technology rather than by DNA sequencing. However, the recently available Conanvarvar software is specialized in identifying very large CNVs of more than one million nucleotides in length.

Bioinformatics Workflow

Figure 4.3 summarizes our bioinformatics workflow.

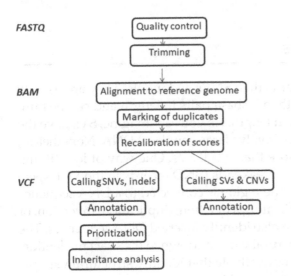

Figure 4.3: Bioinformatics workflow

Summary

In this chapter, we covered the complete process, from extracting DNA to generating a sequence file. We also introduced all the steps of our bioinformatics workflow, from quality control to alignment to the reference genome, calling of the variants, annotation, and inheritance analysis. We also presented the important file formats for genomics analysis: FASTA, FASTQ, BAM, and VCF.

The next three chapters will enable you to master AWS services and Linux, which will prepare you to run the bioinformatics pipeline in Chapter 8.

CHAPTER

5

AWS Services for Genome Analysis

Cloud computing is not a new concept. Ever since data networks have existed, there has been the idea of people here using computing resources there. Way back at the dawn of electronic computing history (the early 1970s), the standard way to do anything with a computer was to sit down at a dumb terminal (i.e., a device that had no computing power of its own but rather transmitted instructions to and received results from a remote computer) and interact with a machine from a distance—usually from somewhere else in the same building.

Later, say from the early 1980s until when the Internet really took off near the turn of the 21st century, we had what was called *client-server* computing. Under this arrangement, workplaces had machines (*servers*) that specialized in particular tasks, such as storing files that many people had to access, running a printer, or hosting a database. People sat at other computers, usually less powerful than the servers, called *clients*. The software on the clients interacted with the software on the servers, and while computational work was done on the client side as well as on the server side, the ultimate goal was usually to update data that resided on a server.

In the case of an office or other workplace, the servers usually resided in a special room in the same building or in another building nearby. Particularly critical servers—say, those that controlled the telephones and contact centers of large companies—were often hosted in data centers, hardened against power loss and other disasters. Users accessed these remote servers by means of some sort of secure network connection. In all cases, however, the servers were 100 percent owned by one entity. If a company decided to deploy a database server, it owned the depreciating hardware, had to pay for its electrical consumption, and had to provide a secure, environmentally controlled place for the server to operate—100 percent of the time. Never mind that the server might be busy between 9 a.m. and 5 p.m. each day and nearly idle overnight or that the server had to be able to handle an extraordinary holiday crush that happened only once a year.

Then, the Internet became generally available, reliable, and fast. That meant people could do useful things with remote computers by means of a browser or other application. In fact, many people could use the remote server (or cluster of servers) simultaneously. Many customers of an airline, for example, could at the same time look at flight offerings and book tickets. Many customers of a bank could simultaneously work with their respective accounts. Many researchers could access a common database as part of their work. However, the remote computing resources were generally in the form of applications, which is to say systems of software that allowed something (such as buying something or sharing information with a social network) to happen.

This somewhat mitigated the problems of owning and operating servers, because they could be made available to more people; anyone with Internet access could use the servers, which allowed usage to be, at least in theory, spread around the clock. Commercial sites could bring in money from a larger potential market, more than offsetting the costs of operating them.

But these were applications that many people could use, not computational resources in their raw form. If you wanted to build an application of your own, you had to provision your own computing power, storage, and everything else you needed. Then, there were major advances in virtual machine technology, in which multiple virtual computers exist on a single hardware platform.

This led to the creation of a new kind of hosted services: cloud computing. Unlike the mainframe computing and client-server architectures that came before and distinct from the hosted applications that remain popular in their niches, cloud computing didn't just allow users to share expensive computing resources at a distance. It allowed the abstraction of computing resources and enabled the implementations of those abstractions to grow and shrink as demand dictated. A particular kind of computing resource—a database server, say, or a place to store files—could be defined, and rather than be constrained by the hardware on which it ran, the resource could grow and shrink as needed.

Perhaps coincidentally, this sea change in computing architecture took place at the same time as the revolution in our understanding of genomics and our ability to manipulate genomic data with software tools. Now, for not a lot of money, almost anyone can build an environment for genome analysis using AWS.

First, you need to understand some tools and concepts.

General Concepts

Before we get into the specifics of the various components you need to build a genomics environment in AWS, you need to understand some basic ideas. Most of them have the potential to be large areas of inquiry unto themselves, but you can approach them in a simplified way for the purposes of what we are attempting to do here.

Networking

Sort of by definition, working with computing resources in the cloud requires some knowledge of computer networking. For the purposes of this book, we'll assume that you are familiar with basic data networking concepts—that you know what an IP address is and what it means for connectivity to exist between computing resources on a network. We won't assume that you've ever worked as a networking specialist, though, and absolutely do not expect you to have any familiarity with the way AWS does things.

AWS Functionalities

AWS provides all kinds of functionalities that allow its users to distribute their resources around the world in an apparently seamless way. This allows AWS customers to improve the durability of their systems (for example, to continue to provide services in the event of a natural disaster that affects a large geographic region) and to offer services of better quality to widely distributed customers (to serve web pages and media files from a geographically proximate location, thereby improving performance). AWS also allows customers to connect its virtual entities to traditional organizational data centers, thereby creating hybrid systems that can meet all kinds of special requirements.

These features generally aren't central to genome analysis, though. For the purposes of what we're doing here, we will assume that you want to do the following:

- Instantiate the tools you need in the AWS cloud as simply as possible
- Access them securely from your own computer
- Use AWS computing resources only as long as required
- Store genomics information more or less permanently on AWS

AWS Accounts

AWS is a business, which means that everything you do with AWS needs to be associated with an account, which is responsible for paying for the AWS services under it. In many cases, you can get quite a bit of value out of AWS for zero cost (at least for a while, as a trial), but you still have to have an account.

To create an account, navigate to

```
portal.aws.amazon.com/billing/signup#/start
```

and work through the process as prompted. You have to provide credit card information for billing, though the first time you create an AWS account, you can use certain resources free of charge—some forever, some only for the first year after the creation of your account, and some for a specified trial period. This is the AWS free tier, and it is good practice to consider what is available in the free tier whenever you want to try a new service.

Virtual Private Cloud

The "stuff" you create on AWS exists on the AWS network, alongside everyone else's stuff. Keeping users' resources separate on a virtualized, shared platform is one of the core responsibilities of the AWS architects and administrators, and their techniques are closely held trade secrets. As far as we are concerned, we can rely on the idea that the AWS resources that belong to us are held within a virtual private cloud (VPC), which is a hosted network to which we control access.

A VPC is an Internet Protocol (IP) network, except rather than taking the form of routers, switches, firewalls, and cables in a rack or data closet, it is virtual. An AWS VPC is a software abstraction of an IP network, which means you can access and make changes to your VPCs remotely. Anything you can do with a traditional IP network can (but need not) be done with a VPC.

In its simplest form, a VPC is a single flat network with a simple address range and no connectivity to other VPCs or the Internet. You can use a simple VPC like this to hook up several virtual servers, called *EC2 instances* (more on them later), the same way you might connect several physical machines to an Ethernet switch with blue Cat 6 cables.

A more complex VPC could have multiple subnets and route tables to describe how resources on each subnet could access resources on other subnets. A still more elaborate one might have an Internet gateway, also with a route table to control access in and out, to let the world at large access the resources on your VPC.

For the purposes of private genomics work, a simple VPC will suffice. You can just create a simple VPC with no capability to interact with the Internet or other VPCs. If you are wondering how you will access computing resources that exist on a basic VPC that does not have Internet access, understand that you will access your AWS virtual machines directly—via encrypted remote access, such as SSH or Remote Desktop. We have a VPC only because it enables virtual machines (and other resources) to share data and also because it's a technical requirement: EC2 instances must exist within VPCs.

To create a VPC, follow these steps:

1. Go to `console.aws.amazon.com`. Establish an account, if you haven't already. Once you have an account, log in using your email address and password.

2. Click Services to see the list of services. Click or search for *VPC*.

3. Click the Launch VPC Wizard button. The first step of the wizard will appear (see Figure 5.1).

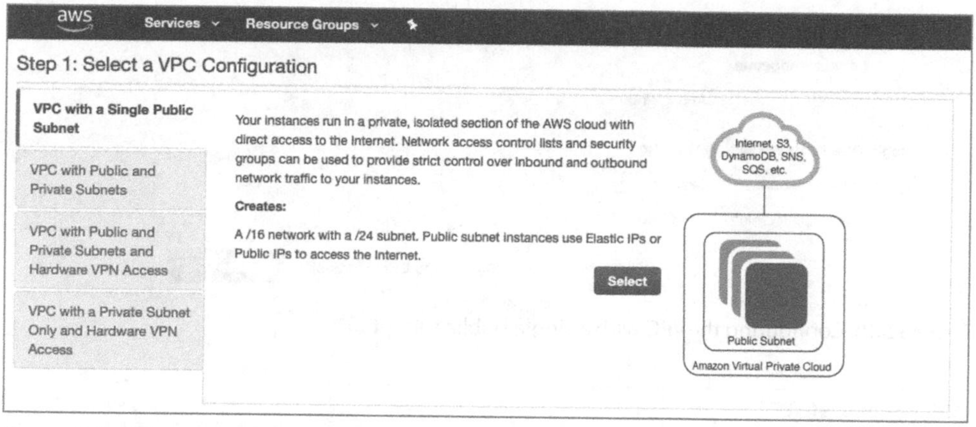

Figure 5.1: Selecting a VPC configuration

4. For purposes of genomics work, choose to create a VPC with a single public subnet. A public subnet is not open to everyone but rather accessible via the Internet by authorized users. Click the Select button. A network configuration page will appear (see Figure 5.2).

5. Specify the IP address characteristics of your VPC. For a small project (meaning one that doesn't need to have a lot of components with IP addresses), even a small network is fine. You can go with 192.168.0.0 and still have more than 250 addresses to use without having to involve a router. Don't forget to give your VPC a name, because AWS will automatically give it a meaningless one if you don't. Click the Create VPC button, and AWS will set up your VPC and display the screen shown in Figure 5.3.

Subnets

Within a VPC, you can create one or more subnets, which are sets of IP addresses that can communicate with one another without the help of a router. (An IP router is a device that moves data packets from one network or subnetwork to another.) Subnets are defined by their subnet masks, which dictate their address ranges.

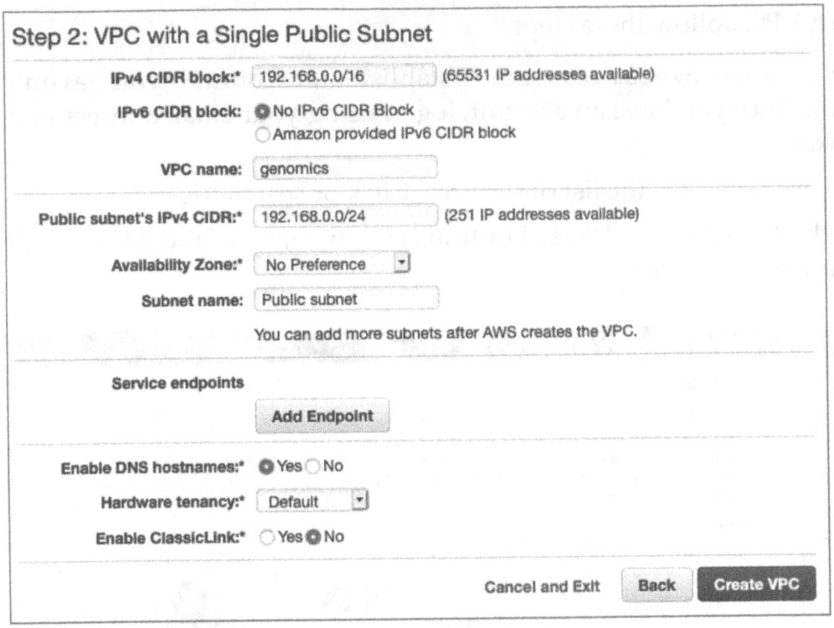

Step 2: VPC with a Single Public Subnet

IPv4 CIDR block:*	192.168.0.0/16 (65531 IP addresses available)
IPv6 CIDR block:	◉ No IPv6 CIDR Block ◯ Amazon provided IPv6 CIDR block
VPC name:	genomics
Public subnet's IPv4 CIDR:*	192.168.0.0/24 (251 IP addresses available)
Availability Zone:*	No Preference ▾
Subnet name:	Public subnet
	You can add more subnets after AWS creates the VPC.
Service endpoints	
	Add Endpoint
Enable DNS hostnames:*	◉ Yes ◯ No
Hardware tenancy:*	Default ▾
Enable ClassicLink:*	◯ Yes ◉ No

Cancel and Exit   Back   **Create VPC**

Figure 5.2: Configuring the VPC with a single public subnet

VPC Successfully Created

Your VPC has been successfully created.
You can launch instances into the subnets of your VPC. For more information, see Launching an Instance
into Your Subnet.

OK

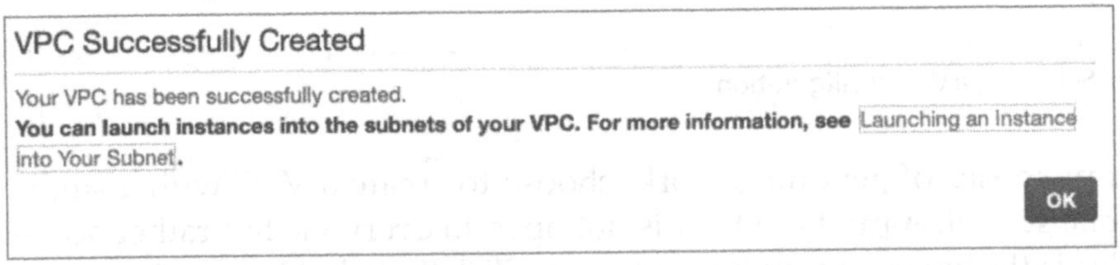

Figure 5.3: Confirming the VPC creation

For example, a subnet could be defined as 192.168.0.0/24, which is the same as saying the 192.168.0.0 network with a subnet mask of 255.255.255.0. That is, half of the 32-bit address space—each of the four numbers separated by dots represents eight bits—defines the subnet, and half is available to address devices on that subnet.

In the case of the 192.168.0.0/24 subnet, you could have 254 addressable hosts—their addresses would be 192.168.0.1 through 192.168.0.254.

It gets much more complex than this if you want to have many subnets and establish communication among them, but for purposes of genomics research, a single subnet is usually enough.

Be aware that 192.168.0.0/24, in which the number of subnet bits appears after the slash, is an example of Classless Inter-Domain Routing (CIDR) notation. AWS usually asks for network addresses in CIDR format.

Elastic IP Addresses

By default, AWS EC2 instances do not have fixed Internet-accessible IP addresses. You can request what's called an *elastic IP address*, which is an Internet-accessible IP address that you can map to an EC2 instance to make it accessible from the Internet. The idea is that you can also quickly re-map an address from one EC2 instance to another (manually or by automatic means), making it possible to recover from failure that way.

The thing to note about Elastic IP addresses is that having them does not attract charges from AWS as long as they are in use. In other words, you pay nothing if you request an Elastic IP address and then assign it to an active EC2 instance. If you ever destroy an EC2 instance, as you typically will when you're done using it, its associated Elastic IP addresses are not automatically released. They sit in your account and you get billed for leaving them idle, so be sure to release Elastic IP addresses you no longer need.

Custom Environments

When building a custom genomics environment on AWS, you need to consider the same components you'd need to think about if you were building a scientific computing system on your own premises. The concepts are the same; it's the implementation that is somewhat different in the age of cloud computing. You need to think about the following components:

- **Computing**: How much processor power is required to do the computational task you want to perform? How much memory do you need? Do you need a dedicated server, or can you save money by using spot instances, or maybe even serverless Lambda functions?

- **Storage**: Genomics projects can be very data intensive, involving hundreds of gigabytes or more of data, and unless you're willing to throw your data away after your analysis is complete, you'll need to store it for a long time. This is why data storage can be one of the costliest parts of an AWS genome analysis project. Fortunately, AWS offers a range of storage options, each with different characteristics with respect to durability, availability, latency, and cost. If you're careful with how you set things up, you can reduce your storage costs quite a bit.

- **Databases and analytics tools**: Much of genome analysis involves the manipulation and comparison of files, but results often can be stored in a database. AWS offers several database solutions.

- **Networking**: Connecting AWS pieces together is pretty straightforward. You generally set up a VPC for your project and hook up components within it. There is more complexity, though, when you need to upload and download data.

- **Workflow management**: Workflow management tools, such as Simple Workflow Service (SWF) and AWS Step Functions, allow you to control a sequence of processes asynchronously.

Storage

Say what you like about genomics data, but there's no way around the fact that it's very big. The genomics files associated with a single project can easily occupy hundreds of terabytes.

AWS provides several levels of storage to cater to either short-term storage, long-term storage, or archival storage.

The short-term storage can be used directly by the EC2 instance and is the most expensive: it comprises the EC2 instance store (typically used for storing temporary buffers and cached data during the lifetime of the instance); the Elastic Block Store (EBS), which corresponds to durable, block-level storage volumes that can be attached to a running instance; and the Elastic File System (EFS), which provides scalable file storage that can be attached to several running instances.

The long-term storage consists of the Amazon Simple Storage Service (S3 for short) and is less expensive than the short-term storage. For cost reasons, the large genomic files such as the BAM files are usually kept on S3. Popular genomics tools such as Samtools or recent releases of the Integrative Genomics Viewer (IGV) can directly work from BAM files stored in S3. However, most other genomics tools require files to be transferred to short-term storage to be processed by the EC2 instance.

For archival needs, data can be stored on Glacier.

When thinking about storage, consider the features that are important to you. These typically include the following:

Durability   *Durability* is a measure of how protected your data is. All S3 storage classes (with the possible exception of S3 One Zone-Infrequent Access) offer the same level of durability (99.999999999 percent). Your data, stored in S3, is unlikely to disappear.

Availability   *Availability* is the amount of time, expressed as a percentage of a specified time period, that your data can be accessed. The S3 Standard storage class advertises availability of 99.99 percent over one year, meaning your data could be unavailable for as long as 52 minutes, 32 seconds, across an entire year without breaking the AWS service level agreement (SLA). In case you're wondering, if AWS does bust an SLA on

S3 availability or anything else covered by an SLA, it will issue you service credits in compensation.

Latency   In terms of storage, *latency* is the amount of time it takes to get access to a stored object. For example, accessing files on your laptop's local solid-state hard drive has low latency. Accessing files on a remote server has higher latency. The S3 storage options have various latency characteristics.

Throughput   *Throughput* is the speed with which a stored object can be accessed, particularly in the content of high-volume situations with lots of requests. It's not normally a concern in genomics workflow applications.

S3

S3 has a number of storage classes, each with different characteristics.

S3 Standard   S3 Standard is the usual storage class. It has the highest cost but also the most ready accessibility. S3 Standard is characterized by high durability, high availability, low latency, and relatively high cost. It is recommended for data that needs to be accessed frequently.

S3 Standard-Infrequent Access   S3 Standard-Infrequent Access (S3 Standard-IA) trades accessibility for cost, while maintaining high durability. In short, slightly more annual downtime is allowed under the SLA, and cost is significantly lower than S3 Standard as long as you access the data less than about once a month. (There are penalty costs if you do.)

S3 One Zone-Infrequent Access   S3 One Zone-Infrequent Access (S3 One Zone-IA) offers functionality similar to S3 Standard-Infrequent Access but costs less and has lower durability. Since it does not redundantly store data across multiple Availability Zones (AZs), data stored using this class will be lost if the AZ in which it's stored is destroyed. This may be a risk worth taking to save some money.

S3 Intelligent-Tiering   There's also S3 Intelligent-Tiering for data the access patterns of which aren't established or which may change over time. Simply put, it will monitor objects and move them between a storage class suited for frequent access and a storage class suited for less frequent access.

When data needs to be accessed less than once a month, consider Glacier.

Glacier

Glacier is Amazon's archiving solution. Unlike S3, which you can think of conceptually as storage on a hard drive on a server, Glacier data is not readily accessible. While Glacier data has high durability (99.999999999 percent, the same as S3), it has relatively very high latency. You can't access your data on a moment's notice but must instead make a request for data and wait for it to become available. In that sense, you can think of it as analogous to a tape archive (which it may or may not actually be).

Within the general Glacier domain, there are two variants of the service. One is for general high-latency storage, and the other is for data that will probably never need to be retrieved.

S3 Glacier   As Amazon's main archival storage service, Glacier is best for data that is very infrequently (perhaps never) accessed. It's the go-to solution for archives and backups. While Glacier is generally the lowest-cost AWS storage option, you can choose latency options (time from request to availability) ranging from minutes to hours, with cost varying accordingly.

S3 Glacier Deep Archive   If you're looking for a purely archival solution, S3 Glacier Deep Archive may be the AWS offering for you. It has high durability, high latency, and very low cost. It's for data you want to keep but hope never to need again.

Computing

Once you have some data in the AWS cloud, you're able to manipulate it. To do that, you'll need some sort of compute element. AWS offers several kinds, including Elastic Compute Cloud (EC2), Elastic Container Service (ECS), and serverless Lambda functions.

Elastic Compute Cloud

EC2 is the virtual server solution on AWS. You can build a virtual machine to your own specifications, either by creating it from the ground up, by using a stock image that AWS makes available (and perhaps modifying it), or by using an image provided by a third-party vendor in the AWS Marketplace. You'll learn how to create an EC2 instance in the next chapter.

Once you have an instance, you have several options with respect to how to run it. Some options favor availability over cost, whereas others can be more economical if you're not as concerned with when your computing tasks run.

The four basic availability options are as follows:

On-Demand Instances   You create an instance and launch it (set it running) whenever you want. You pay by the second for running time. This option is optimized for availability—it's there for you to use whenever you want—but costs the most. You can think of an on-demand instance as your own server sitting in the AWS cloud, waiting for you or your users to connect to it and make use of the applications running on it—the same as if you were connecting to a server in a machine room down the hall. You can turn the on-demand instance on and off via the AWS console whenever you want, and you should have it on only when needed, because again you're billed for "on" time. Even with the server shut down, though, you will pay for any EBS volumes that may be associated with the on-demand instance. You have to delete those when you're done with the on-demand server.

Reserved Instances   If you can commit to a longer period (from a year to three years), AWS offers a discount on what are otherwise the same as on-demand instances. Subcategories of reserved instances include standard reserved instances (which are suited for always-on reserved instances and are the most deeply discounted kind) and convertible reserved instances (which are functionally similar to standard reserved instances, except that they allow instance types to be modified or upgraded over time).

Scheduled Instances   Also called scheduled reserved instances, these instances go on- and offline according to a schedule you configure. Scheduled instances are not as deeply discounted on a per-second basis as standard and convertible reserved instances, which are meant for full-time operation. On the other hand, they allow for operating schedules other than full time—for example, you might set up a scheduled reserved instance to run for only a few hours every day to do a batch job. You have to commit to a full year of a specified schedule to get the instances at a discount. Instances like this are handy for regular, automated tasks, such as doing a nightly summary of transaction data. They are less useful for scientific work, such as genome analysis, which often requires interactive and varied work under the control of a researcher.

Spot Instances   A spot instance is one that comes to life only when AWS capacity availability is such that the company can provide the kind and quantity of EC2 instances you need at a particular price per hour. This is where cloud computing really shines, because it limits waste—EC2 instances and the computing power to run them, which would have otherwise sat idle, can be put to use on jobs that are less time-critical. When AWS has extra computing power, it spins up some spot instances, enabling the jobs running on those instances to move ahead. When there is an increased requirement for on-demand computing—people with reserved instances begin turning on their VMs or it is time for scheduled reserved instances to start up—spot instances can be shut down. To compensate for this on-and-off behavior and the fact that there is no guarantee that an instance will be running at a given moment, users of spot instances are given deep discounts—70 percent or more below on-demand instance prices. This is the lowest-cost option of all the EC2 run modes and is suitable for the kinds of batch processing that genomics workflows sometimes require.

Aside from the way in which they run and are priced, EC2 instances are grouped into instance types, which is to say the applications for which they are generally suited. The following instance types are available:

- General Purpose, which have names beginning with M, or sometimes T or A (we think of M for "miscellaneous" computing). These machines have a balance of computing power, memory, and network performance (don't forget that network performance governs how an instance accesses remote objects, such as those in S3 buckets).

- Compute Optimized, which have names beginning with C (presumably for "compute"). These machines have lots of processing power.

- Memory Optimized, whose names start with R or X (think RAM or "Xtra" memory). You might consider these VMs if your planned application includes rapid and random access to stored objects, and putting those objects in persistent storage would slow things down too much.

- Accelerated Computing, which have names starting with P, G, or F. These virtual workstations have graphics processing units (GPUs) or field programmable gate arrays (FPGAs) of the sort that can be used for computationally intensive work—we associate G with GPU and F with FPGA.

- Storage Optimized, which have names beginning with I, D, or H (think of I for IOPSs, which are input-output operations). Use these machines for work that involves a lot of accessing stored objects.

A typical EC2 instance type name would be something like m5.large, which describes a general-purpose instance (the fifth version thereof) of large size (other size options include xlarge, 2xlarge, and so on). You can look up the exact specifications of this or any other instance type in the AWS EC2 documentation and discover that the m5.large has two virtual CPU (vCPU) cores, 8 GB of RAM, and support for EBS storage. These instance specifications evolve all the time, so be sure to consult the documentation for the latest information.

Containers

Virtual machines, which is to say EC2 instances in the AWS world, are not the only way to package up software solutions and run them in the cloud. Containers are a means by which application software, including all of its dependencies, can be stored in a single package and run anywhere. Typically, they are run on a developer's workstation initially and then deployed to much more powerful computing resources when it is time to put them into production.

Think of containers as highly portable applications. In a single module, containers hold whole software solutions, complete with all their dependencies.

There's no operating system within a container, so they are much "lighter" than full virtual machines.

Containers are not unique to AWS—they are a generic technology that can be used in a variety of ways. The central technology of modern containerized software is (as you would expect) the container, which is ultimately a file format. Chief among container designs is Docker, an open-source specification for virtualization and the operating system level. (Docker is not the only container specification, and it is in fact based on other specifications like LXC and runC, but it is the most widely used.) To be more precise, Docker is both a specification and a company—Docker, Inc.—that offers products (all with variants of the Docker name) that implement the Docker standard. Some of its products are free; others have costs associated with them. It can be more than a little confusing.

The Docker standard allows a single OS instance—a single Linux server, to cite a simple case—to host multiple Docker containers. Each of the containers is an insular world unto itself, aware of the computer on which it is running but with no perception of any other containers. Note that the containers are not virtual machines, because they do not contain the OS or any virtualized CPU or RAM resources, the way a VMware or EC2 instance does. You can say that containers are virtualized at the OS level, rather than at the hardware level. This makes them smaller, more portable, and generally more efficient.

On the other hand, containers don't have guaranteed resources, the way virtual machines do. If you set up a VM, say in EC2, and say that it has 8 GB of RAM, that is how much RAM it will have. If you put several containers on a host and that host has however much RAM it physically has, then the containers need to share that physically limited resource. An important trade-off for the "lightness," portability, and simplicity of containers is the lack of guaranteed resources.

Once you have containerized software, you need a way to run it. In the simplest case, on a single physical server, you might set up a series of Docker containers—for example, one for a web server, one for business logic, and one for database work in a multitier application. You then need a way to run these containers and have them talk to each other in a predictable and controllable way. This is the function of the Docker Engine (a product of the Docker company, in both free and paid versions), which acts as the master process to coordinate the running of containers. That's how it works on a single server, and it is probably how you will set up your development machine.

Things become more complex when you need to coordinate the running of containerized software across multiple servers—a cluster. Such coordination requires taking multiple pieces of hardware, perhaps with disparate capabilities, and managing the operation of multiple containers across them. Load will be spread across all available computing resources in such a way that no item of hardware is overworked, and all containers enjoy the highest possible performance. There needs to be a way to simulate IP networking, with configurable rules about what containers can communicate with what other containers and how.

Still staying clear of cloud offerings for now, there are a couple of packages that facilitate this requirement.

The Docker solution for the coordination of containers across a cluster is called Docker Swarm. (Again, it is an open-source solution maintained by the for-profit Docker company.) Swarm, as it is often called, has a reputation for being simple and easy to configure. It allows users to use the YAML definition language to group containers into units called *services* or *microservices*, which usually correspond to functional applications (interface, business logic, and database, for example).

The main alternative to Docker Swarm is Kubernetes, which was developed by a group of ex-Google software engineers who had worked on Google's internal container management solution, known as Borg. Kubernetes was developed and subsequently released to the public as open-source software. Kubernetes can manage Docker containers, as well as other kinds, and allows its users a great deal of flexibility in determining how their containers are grouped, both functionally and with respect to their physical allocation across a cluster of servers. A pod is a group of containers that are colocated on a single server within the cluster, and a service is a functional entity that spans pods to provide some desired capability. Beyond that, Kubernetes allows for a common filesystem to be shared across multiple pods in a more or less transparent way.

Are you confused yet? With some variations with respect to the specifics of implementation, containerization allows functional computing units to be separated from their operating systems and treated as discrete entities. These entities can then be grouped and made to run in a coordinated way across multiple instances of the operating system—multiple physical servers.

Now, consider how this can be done in the cloud, where everything is generally virtual and scalable.

The most straightforward way to deploy and run containers under AWS is with Elastic Container Service. You can register a container with ECS and then use it to run the containerized applications and make them available to users. You can use the management service to scale your application up and down according to demand.

It is important to understand that under ECS, the concept of EC2 virtual machines does not go away. ECS manages containers that run on one or more EC2 instances. Recall what we said a few paragraphs ago, about Docker containers running (singly or in groups) on multiple Linux servers and something—such as Docker Swarm or Kubernetes—managing the operation of the collective resource. Under the AWS architecture, ECS assumes a role parallel to that of Swarm or Kubernetes. It manages containers running on multiple server instances. You can fit other resources, such as load balancers and EBS volumes, into your ECS configurations, as well. It is worth noting that ECS attracts no charges of its own. Any costs associated with an ECS configuration are those associated with subsidiary resources, such as the EC2 instances and EBS storage.

The second container management solution available under AWS is Elastic Kubernetes Service (EKS). As you might imagine, EKS is the AWS way of offering Kubernetes container-management software. Similar to the way in which you would set up ECS, you set up an EKS environment (you will see it called a *control plane* in a lot of the documentation) that runs on worker nodes, which are themselves EC2 instances. You then associate the EKS control system with EC2 instances, which you also need to set up. Your Docker containers then run on those EC2 instances, with EKS managing load and dealing with such eventualities as node failure. As is the case with Swarm versus Kubernetes in general, the decision to use ECS or EKS comes down to the demands you anticipate placing upon your distributed container architecture.

Think of ECS and EKS as roll-your-own container management solutions under AWS. If you want a solution that provides you with more guidance, in exchange for decreased flexibility or configuration and increased cost, you should consider AWS Fargate. Fargate handles the configuration of a container-running cluster in a much more automated way, taking a certain amount of control away from you but allowing swift deployment of elaborate container systems. Fargate abstracts servers altogether, meaning there is no need to think about the EC2 instances that underlie a container-running cluster at all. It is similar to Lambda in that your code (in containers) runs in an apparently serverless way. While of course there are servers running your containers, you cannot see them and do not have to think about how they are configured.

One pays for running containers under Fargate in a way similar to that of Lambda functions. It is all based on usage, with one price per vCPU per second and a second price per gigabyte of RAM per second. (You will need some of each.) All times have a one-minute minimum, and you can gain discounts by committing to a certain level of usage in advance.

For genomics work, Fargate is probably your best bet. Your containerized workloads will not be running for very long, the way they would be if your application concerned some business function that had to be left running and available to users for long periods of time. For that reason, the slightly higher cost of Fargate will not amount to much in absolute dollar terms. You can build your Docker containers on your local workstation, upload them to AWS, and set up a Fargate application quickly. It is a good option if you want to save time in configuring containerized applications, enabling you to focus more on genomics work.

Lambda Functions

You can think of Lambda functions as much smaller versions of containerized software configured and run by Fargate. They are similar to Fargate applications in that they run on virtualized servers, meaning that the servers are not "visible" to us when we make use of them. The servers are abstracted away in the AWS environment. Developers who make use of Lambda functions do not have to worry about deploying a server to host them. The Lambda function is just deployed into the AWS environment, is maintained as an available resource that is part of your AWS infrastructure, and can be called as needed.

Lambda functions are essentially executable modules of code (you can write them in Python, Node.js, C#, and Ruby, as well as Java, Go, and Windows PowerShell) that can be called from anywhere, security configurations permitting. A lot of things can invoke Lambda functions, and perhaps most often they are called in response to a triggering event. For example, an S3 bucket can trigger a Lambda function when a file is written or changed. Alternatively, a Lambda function can run on a time schedule or be exposed for larger applications to call by means of an API gateway. With API gateways, you can build complex applications that are run in a distributed way, with Lambda functions called by (for example) mobile apps, web pages, and traditional compiled desktop applications.

The cost of Lambda functions is on a per-invocation basis, so there's no need to maintain a server that might be substantially idle much of the time.

Workflow Management

To coordinate multiple steps running asynchronously, you need a workflow management solution. AWS has three: AWS Batch, Step Functions, and the older Simple Workflow Service (SWF).

AWS Batch

AWS Batch allows you to do the same things you can do with Linux shell scripts (.sh) and Microsoft Windows batch (.bat) files, which is to say run processing jobs in coordinated groups using (for example) the output of one job as the input of the next job.

Based on Docker containers, Batch coordinates the running of jobs on EC2 instances and can even optimize tasks to take advantage of lower spot pricing; you can specify that Batch runs a job only when computing power is available at or below a price threshold you specify. AWS Batch jobs cost nothing in themselves. You pay only for the computing and other resources you control by means of AWS Batch.

AWS Step Functions

Amazon Step Functions are concerned with sequences of tasks that must be performed asynchronously. That is, the actual execution of steps is carried out by different resources—perhaps a piece of software on an EC2 instance here, perhaps a Lambda function there, then a call to a RESTful web service, then a batch script, then maybe a human action of some kind. While workflows need not be strictly linear, there's generally a requirement that one step completes before the next step begins—the output of one becomes the input of the next. When it is impossible to predict how long it will take a given step to complete, we need something to coordinate the overall process. That's where Step Functions come into their own.

In genomics work, Step Functions can manage the overall process of taking a FASTQ file, transmitting it to a compute resource for analysis, and taking the results—a BAM file, typically—and storing them somewhere else. It can also (as well or instead) store the results of the sequence analysis in a database for further examination that way.

Simple Workflow Service

Functionally similar to AWS Step Functions, AWS SWF requires the user to develop "decider" code modules. Although they are more flexible than workflows implemented in AWS Step Functions, SWF solutions can be harder to develop, test, and maintain. AWS recommends that you use AWS Step Functions unless you have a reason not to.

Third-Party Solutions

The Broad Institute has created a scientific-oriented workflow execution system called *Cromwell*. Cromwell is based on AWS Batch and uses a dedicated control server to run Batch jobs that coordinate the activities of multiple worker nodes and the flow of input and output files between the worker nodes and an S3 storage bucket.

The Barcelona Centre for Genomic Regulation developed a workflow framework named *Nextflow*. Nextflow is functionally similar to Cromwell and also makes use of AWS Batch.

Summary

In this chapter, we explained the value of cloud computing in general and how it differs from both client-server technologies and hosted applications. We showed that the AWS cloud provides tools and makes them available for use by many people simultaneously, which has the double advantage of giving people access to computing resources they might otherwise not be able to afford and of using shared computing resources efficiently.

We discussed some specific AWS services that are relevant to genomics work—most notably EC2 instances, which are the virtual computers that run genomics software, and S3 buckets, which are the massively scalable storage containers suitable for holding large data sets securely and reliably. We explained how to choose an EC2 instance type that suits the work you want to do and how to configure it to run—whether on-demand, on a schedule, or according to a spot price to reduce cost. We also wrote about the various features of S3 and how it can be optimized for speed, redundancy, or cost—while maintaining high reliability and availability regardless of that optimization. We also explained Docker containers and showed the various AWS tools you can use to configure and run them.

In the next chapter, you'll see in detail how to set up an AWS environment for genomics work.

# CHAPTER 6

Building Your Environment in the AWS Cloud

In this chapter, we'll walk through the process of setting up resources in three AWS services that any genomics project will require: Virtual Private Cloud networking, Elastic Compute Cloud compute power, and Super Simple Storage.

Setting Up a Virtual Private Cloud

To have Elastic Compute Cloud instances in the AWS cloud and have them connect to anything, we need to establish a virtual private cloud (VPC). As you learned in Chapter 5, a VPC is a network inside the AWS infrastructure. You can think of it as a sort of virtual Internet Protocol (IP) network, similar to what you probably have at home and at work (the Ethernet and Wi-Fi networks that connect computers and other devices). Like physical IP networks, VPCs can include subnets (with which network-connected devices are separated from each other by means of their IP addresses), routers and firewalls (which connect subnets to one another in configurable ways, and gateways (which allow an IP network to communicate with other networks, usually including the Internet).

There is a lot to be written and understood about VPCs, but most of it does not apply to the work of genomics researchers. All we need to do is set up a simple VPC with a single subnet to host whatever AWS computing resources we need to perform our genomics work.

To set up a VPC, log into the AWS Console and navigate to the Virtual Private Cloud service, either by choosing it from the (giant) list of options that appear in the Services drop-down menu or (more easily) by entering **VPC** into the Find Services search box. Figure 6.1 shows the main console page.

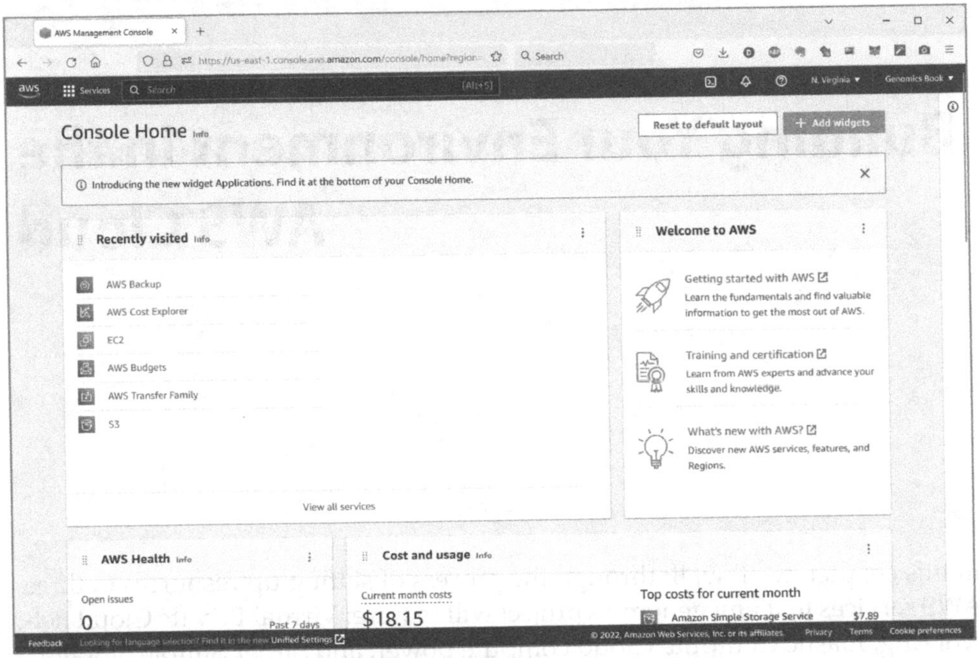

Figure 6.1: The main console page, where you should search for VPC

On the main VPC page, shown in Figure 6.2, you will see information about existing VPCs and other network resources (such as gateways) that exist in your account. Even if you have just created a new account, there are some resources that exist by default.

The easiest way to create a new VPC—and the correct way to do it if you're setting up a straightforward genomics work environment—is to use the VPC Wizard. It will ask what you want and set up a VPC to meet your requirements. The VPC Wizard is able to handle situations far more complex than ours, so click the Launch VPC Wizard button to start the process.

Step 1 of the VPC Wizard, shown in Figure 6.3, asks you to specify what kind of VPC you want.

Two of the options have to do with connecting a noncloud network to the AWS VPC by means of a virtual private network (VPN). That's something you would do if you wanted to allow users of a physical IP network (in an office, for example) to connect to resources on AWS, but it's not what we want to do.

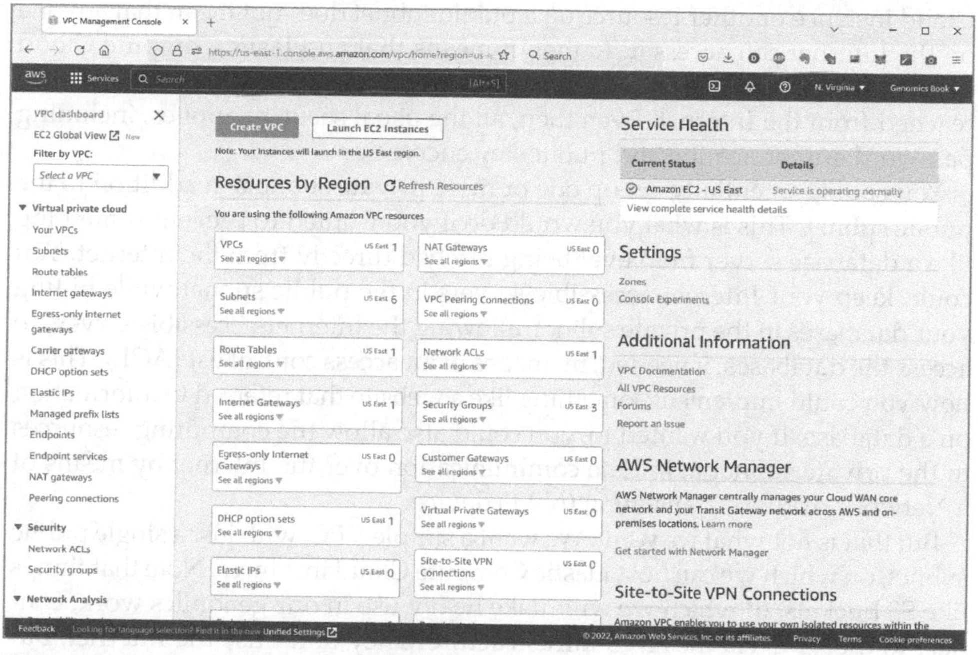

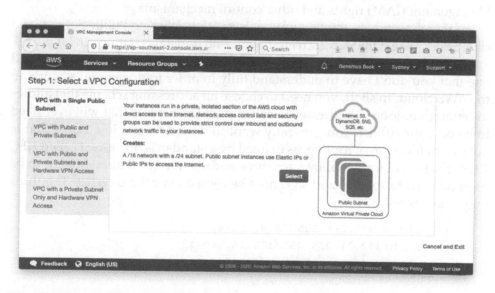

Figure 6.2: Existing VPCs and other network resources

Figure 6.3: The initial step of the VPC wizard

The other two options (the first two listed) have to do with subnetting. You need a public subnet to access your AWS Elastic Compute Cloud instances over the Internet, so both options offer that. Note that having an Elastic Compute

Cloud instance or other resource on a public subnet does not mean that anyone on the Internet can access it. Rather, it means that an Elastic Compute Cloud instance, if it is given an Elastic IP address (or other public IP address), can be reached from the Internet. Even then, all the usual security applies, including password authentication and public-key encryption.

You could, potentially, set up one or more private subnets in addition to the public subnet. This is what you would do if you wanted to prevent something like a database server from ever being reached directly from the Internet. You could keep your Internet-accessible servers in the public subnet while hiding your databases in the private subnet, allowing the Internet-accessible servers to access the databases, if needed, by means of an access control list (ACL). This is how you could implement something like a website that referred to information on a database. If you wanted to, you could also allow the computing resources in the private subnet to initiate communication over the Internet by means of a Network Address Translation (NAT) device.

But that is not what we want. We want a simple VPC, with just a single public subnet, on which we can host Elastic Compute Cloud instances. Note that things like S3 buckets, of which we will make heavy use in our genomics work, connect to the VPC via the AWS infrastructure. They don't use the Internet, but rather AWS connectivity that is controlled by means of AWS Identity and Access Management (IAM) rights and other control mechanisms.

We therefore choose the first option, VPC With A Single Public Subnet, and click Select. We get the Step 2 screen shown, with its default settings, in Figure 6.4.

IP addressing schemes are a large and somewhat complicated subject, and one that you don't have to understand fully in order to do genomics work on the AWS cloud. In short, you need to choose an addressing scheme that allows for enough unique IP addresses to cover all of the devices that will ever need to be on your subnet. You generally want to use addresses in a range that is not meant for Internet routing, as defined by a standards document called RFC 1918. That specification allows for three address ranges that are meant to be used on private subnets and so cannot be routed over the Internet. The three ranges are as follows:

```
10.0.0.0 to 10.255.255.255 (10.0.0.0/8)
172.16.0.0 to 172.31.255.255 (172.16.0.0/12)
192.168.0.0 to 192.168.255.255 (192.168.0.0/16)
```

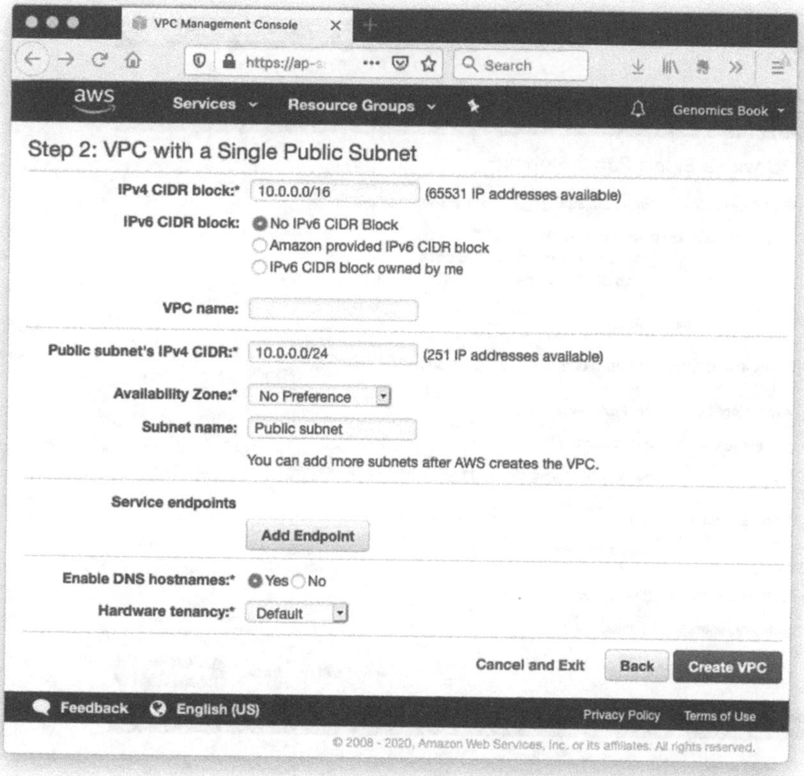

Figure 6.4: Specifying network characteristics of the new VPC in the VPC Wizard

Without getting into the nuances of how subnetting works—search for *subnet masking* if you want to understand—you can see that the three ranges have different numbers of unique hosts. For the purposes of the small collection of machines that will live on our VPC, we will use a large subnet mask. The subnet 192.168.0.0/28 allows for 11 unique addresses, which should be plenty for our purposes. (AWS requires a subnet mask between 16 and 28 anyway.)

After we specify the same subnet as our public subnet's IPv4 CIDR and give our VPC a name, we have a configuration box that looks like what is shown in Figure 6.5.

Click the Create VPC button to build your VPC. AWS confirms that it has been created, as shown in Figure 6.6, and the new VPC appears as available in the list of your VPC, as shown in Figure 6.7.

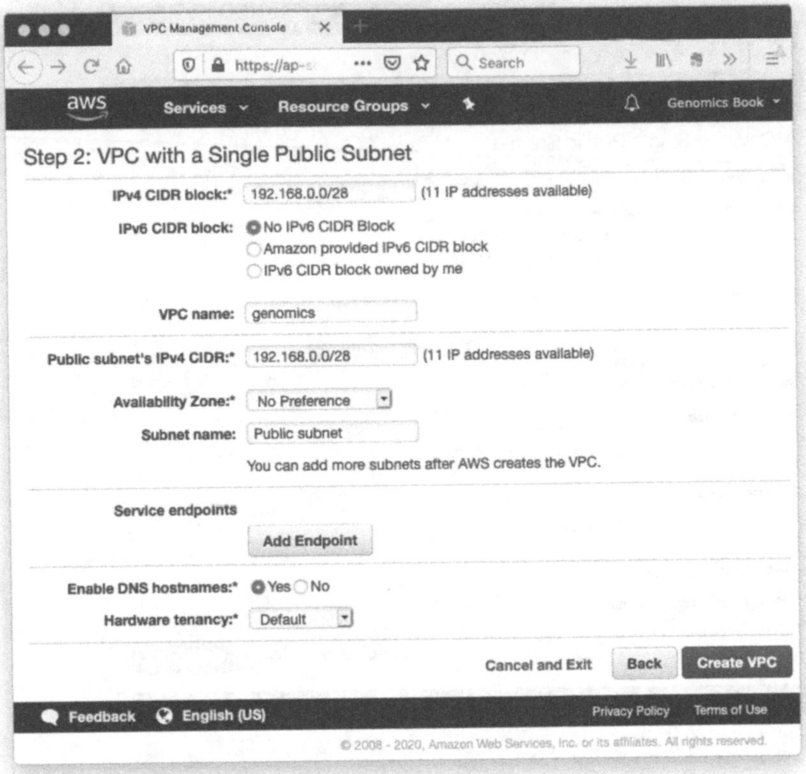

Figure 6.5: The second step of the VPC wizard, with VPC name and address block specified

Setting Up and Launching an EC2 Instance

We will use the Elastic Compute Cloud (EC2) service to create and run servers that will do our genomics work.

To set up an EC2 instance—a virtual machine, in other words—perform the following steps:

1. Log into the AWS Console and navigate to the EC2 service, either by choosing it from the Services drop-down menu or by entering **EC2** in the Find Services search box. Figure 6.8 shows the main EC2 status page.

2. Click the Launch Instance button to start the process of configuring a new EC2 virtual machine. The first step, shown in Figure 6.9, is to select an image to use as a basis for building the new instance.

3. We like to do genomics work under Ubuntu Linux, so search for *Ubuntu* in the search bar and click the Free Tier Only box so that we can do our work as economically as possible. The results appear in Figure 6.10.

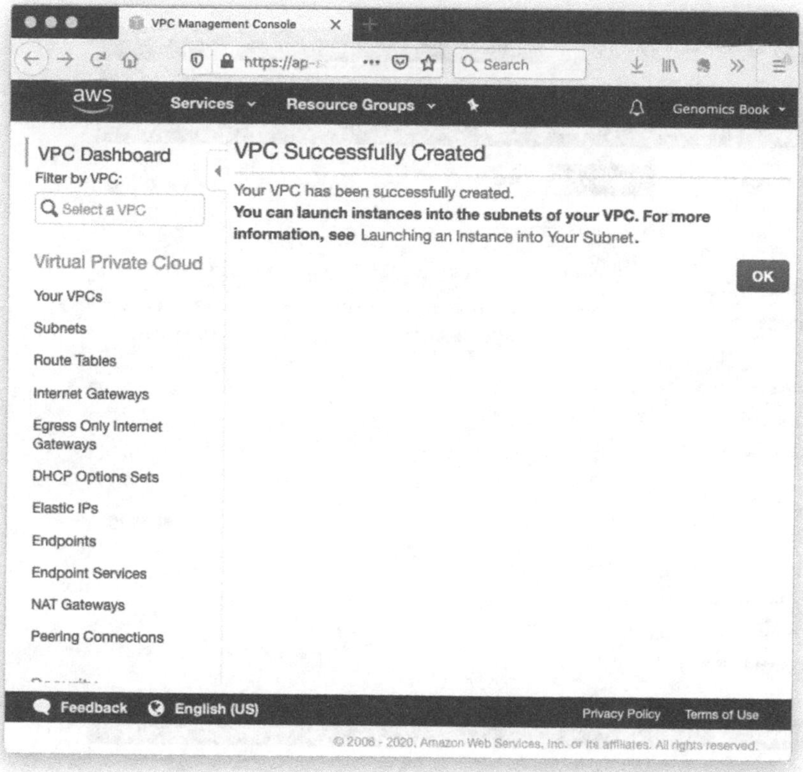

Figure 6.6: VPC creation success

4. Ubuntu Server 18.04 suits us fine, so click the Select button to the right of that option.

5. The next step is to specify the virtual hardware on which our new instance will run, using the list shown in Figure 6.11. There's no way to filter for free tier eligibility here, but there's only one option anyway: a small, general-purpose configuration called t2.micro. We will choose that.

6. From there, jump right to the conclusion with the default settings by clicking the Review And Launch button.

   However, we'll go through the rest of the steps so that we understand everything.

7. Click the Next: Configure Instance Details button, and see what appears in Figure 6.12.

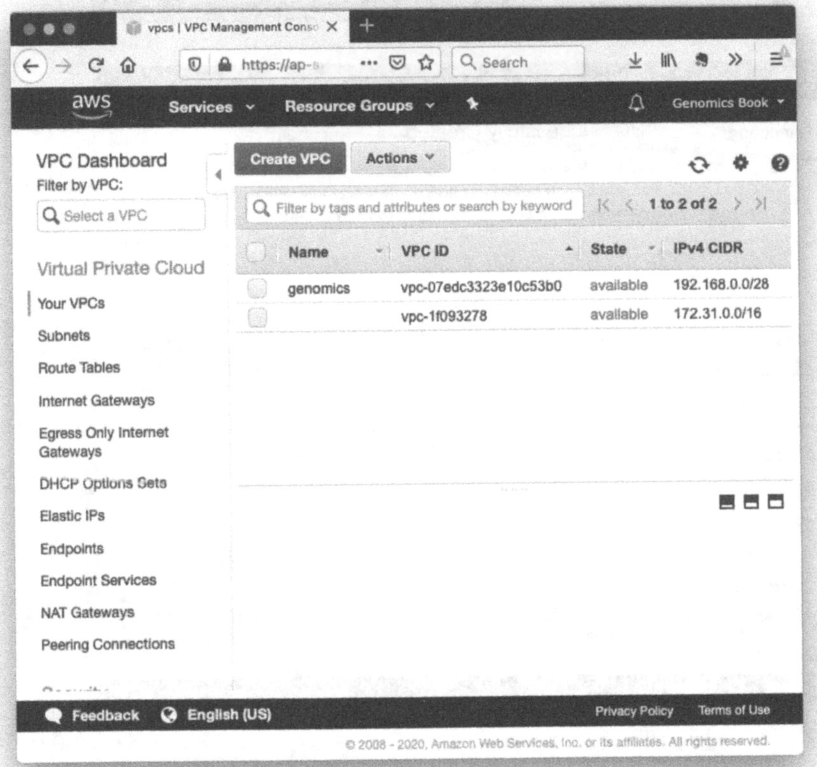

Figure 6.7: The newly created VPC among other VPCs

The only change required here is to choose your VPC from the drop-down list box next to Network. You also have to decide how you want to get an IP address for the instance you are creating. (It will need a public IP address if you want to access the EC2 instance across the Internet). You can either choose to have a public IP address assigned each time the instance starts up or assign an Elastic IP address to the instance (something you would have to set up separately). Using an Elastic IP address would probably be the way to go if you want your instance to have a consistent address—for example, if you were building a web server whose IP address had to be resolvable by the global Domain Name Service (DNS). That's not the sort of thing we are building here. We will be the only ones using this virtual computer, and we would rather deal with getting a new address every time it starts up than pay for an Elastic IP. We will therefore choose Enable in the box next to Auto-assign Public IP and then click Next: Add Storage.

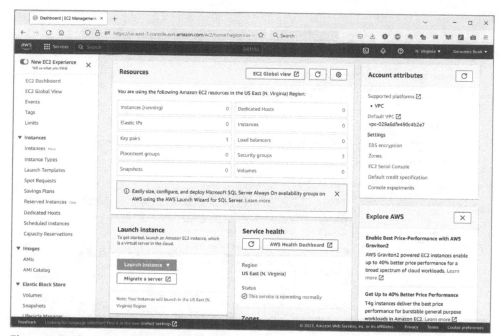

Figure 6.8: The main EC2 status page

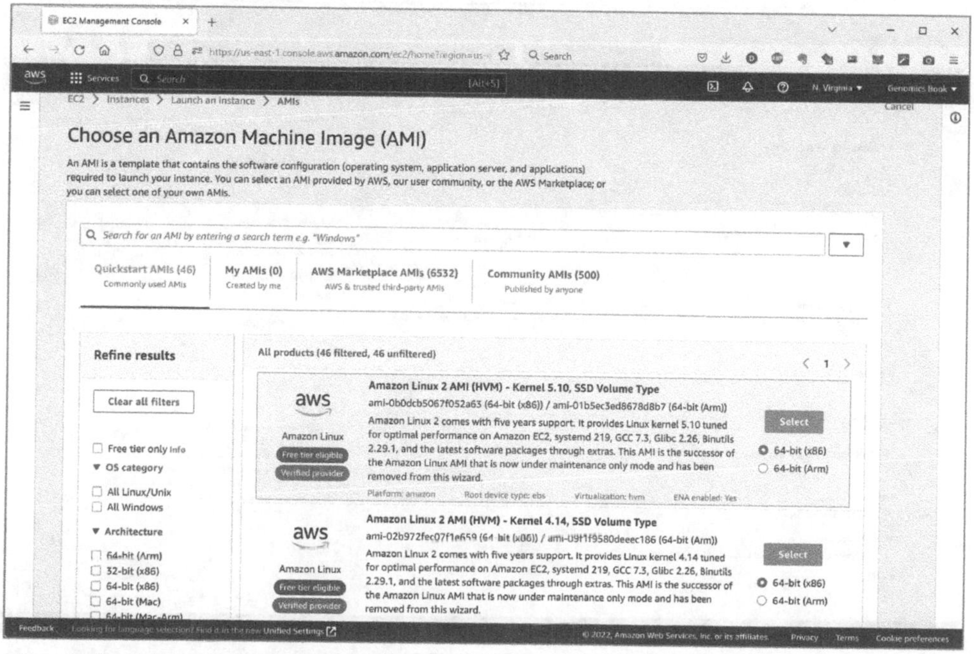

Figure 6.9: Selecting an image for the EC2 instance

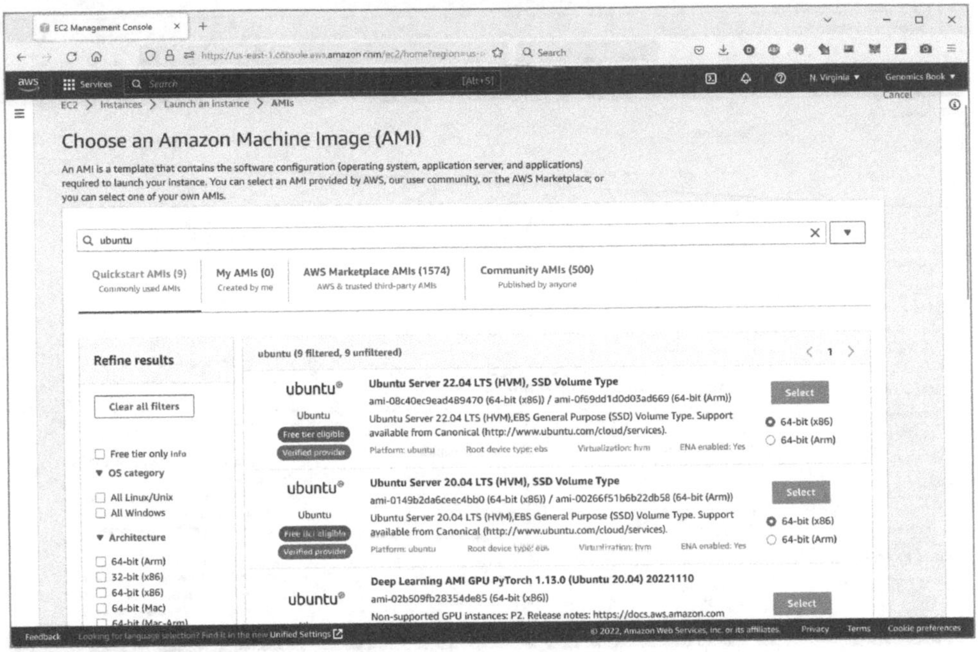

Figure 6.10: Choosing an AMI image to use as a base for the new server

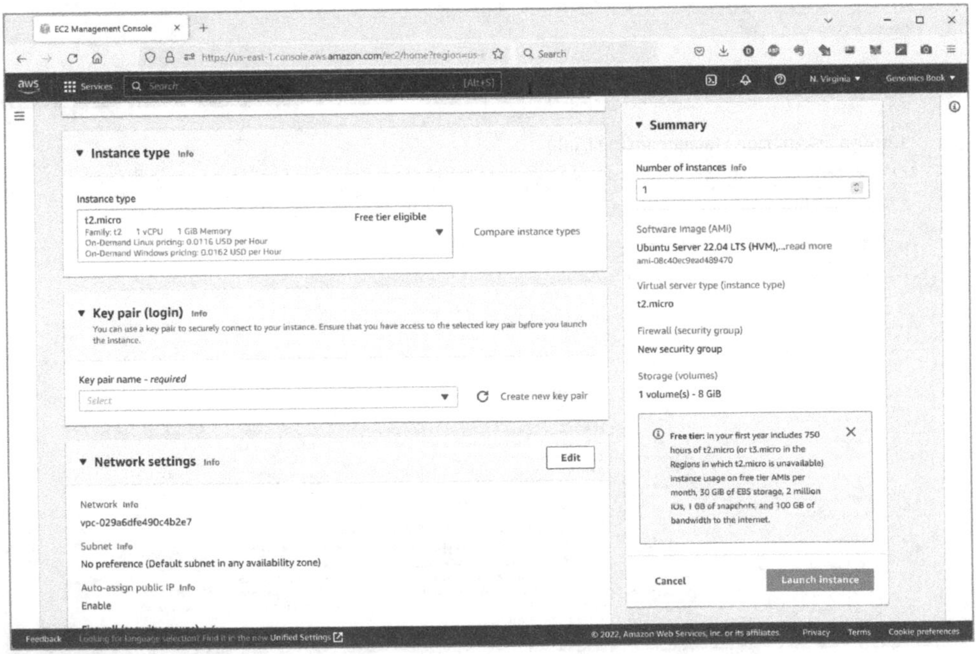

Figure 6.11: Specifying the virtual hardware for the EC2 instance

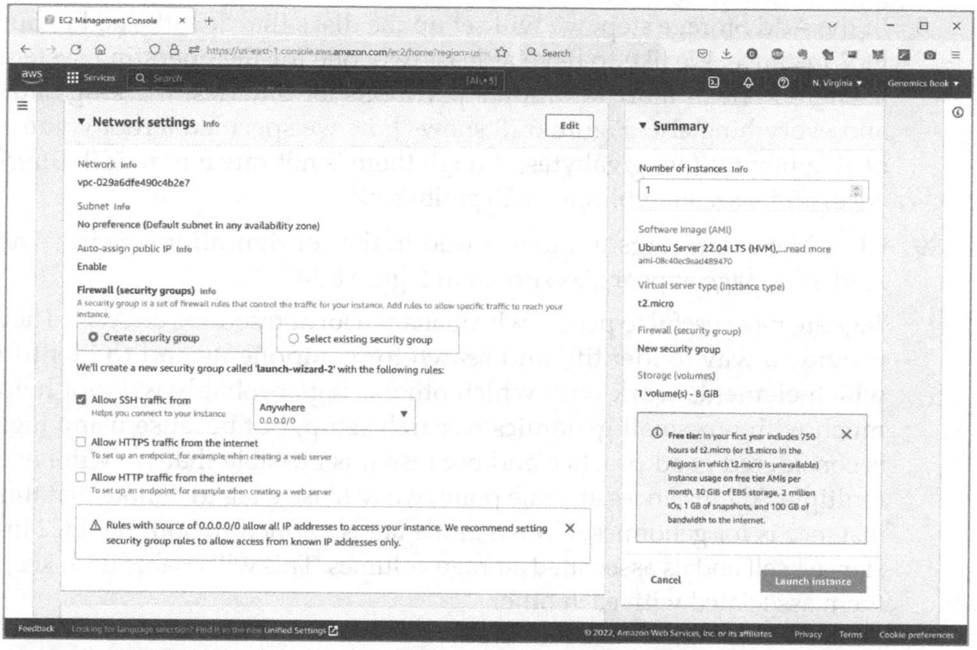

Figure 6.12: Attaching the new EC2 instance to the VPC created earlier

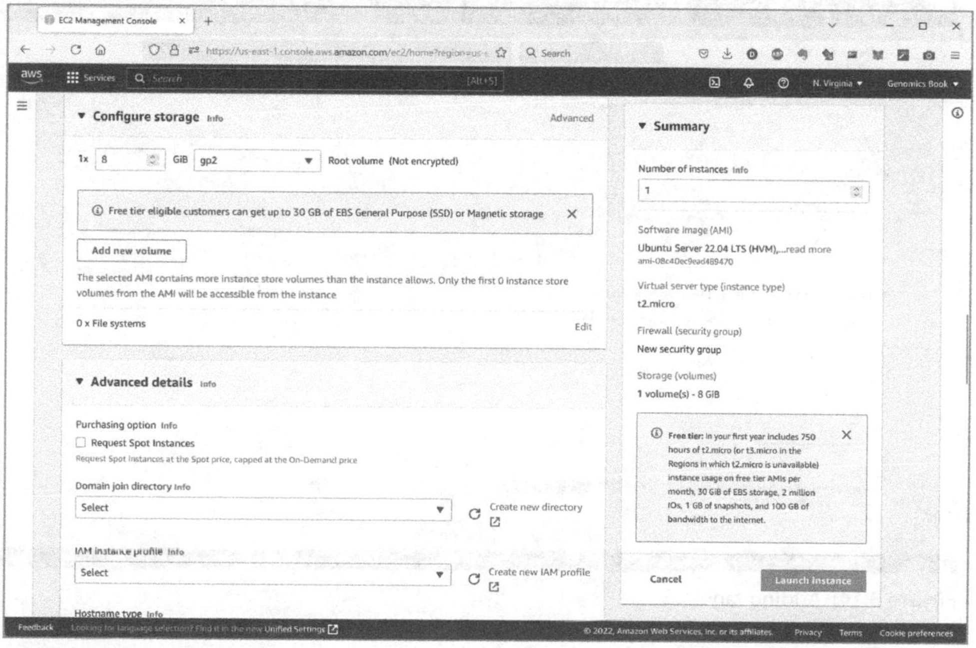

Figure 6.13: Adding storage

8. In the Add Storage step, we will set up the disks that are attached to our EC2 instance. We like to have at least two: one for the operating system itself and one or more additional partitions for binaries, working data, and everything else. Figure 6.13 shows how we specified a root volume of 10 gibibytes (not gigabytes, though there is not much practical difference) and a second volume of 20 gibibytes.

9. Click Next: Add Tags to move ahead in the configuration wizard. The Add Tags page appears, as shown in Figure 6.14.

   Tags are most useful to people who manage a lot of resources on AWS. They provide a way to identify and search for components and to identify which elements work with which others. Tags probably will not help much with our small genomics research setup, but because using tags is considered good practice and because it is possible that we will have multiple EC2 instances at some point, we will use a tag to denote that this instance is for genomics. Note that the tag is applied to both the EC2 instance itself and its associated storage volumes. This will enable us to keep them associated with each other.

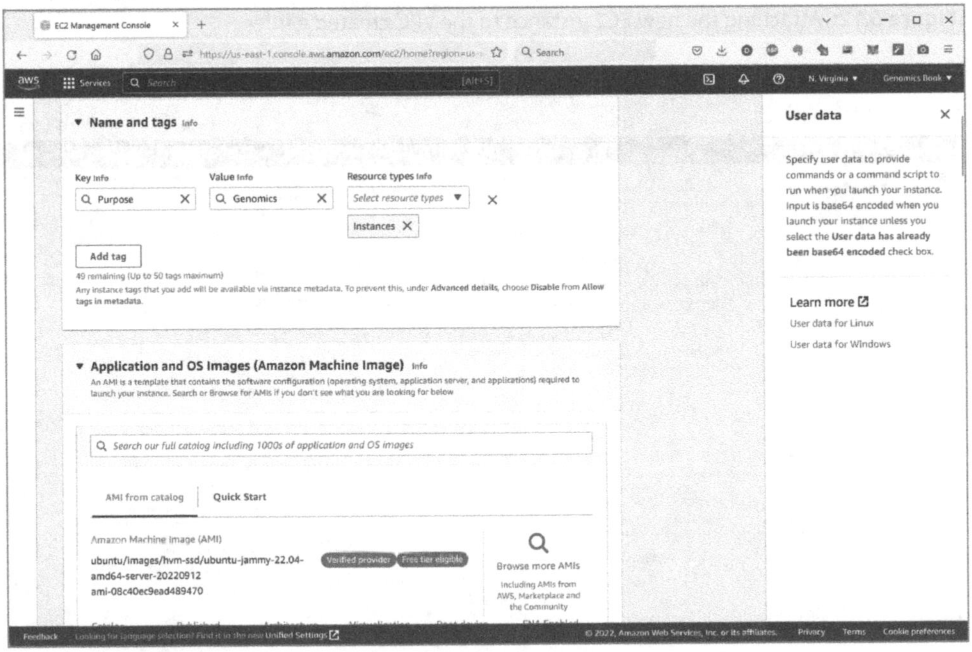

Figure 6.14: Adding tags

10. Click Next: Configure Security Group to move on to the next step.

    A security group is essentially a firewall that determines what Internet hosts will be allowed to access the EC2 instance, as well as the protocols they will be permitted to use, and on which ports. It is considered good practice to limit access as much as possible. In this case, we will allow access only via SSH on port 22 from the IP address of our home and also from one of our workplaces. Figure 6.15 shows how we set this up.

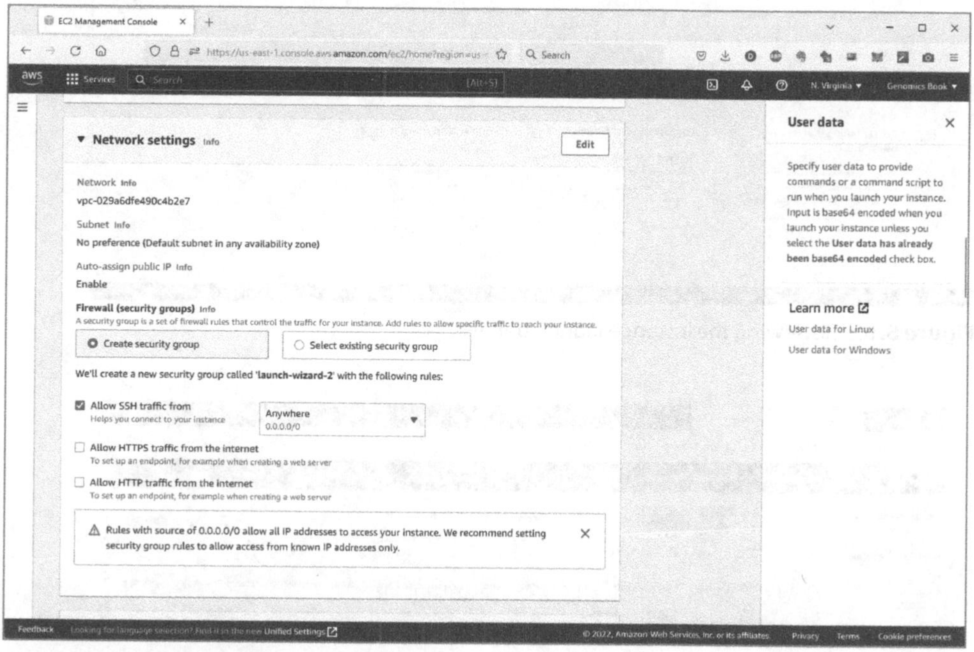

Figure 6.15: Configuring a security group

11. Click Review And Launch to build the EC2 instance. You will see a final summary page, as shown in Figure 6.16. Make sure everything is as you want it, and then click Launch.

12. Before your instance builds, though, AWS will prompt you to specify the encryption keys that protect the traffic into and out of your instance. This takes the form of a public key (kept by AWS) and a private key (which you download and keep in the form of a .pem file). For the purposes of this exercise, specify that you want to create a new key pair, and give it an intuitive name, before downloading the .pem file.

The new instance will immediately appear in your list of EC2 instances but will be shown in the "pending" state until it is ready, as shown in Figure 6.17. When the instance state turns to "running," it is available for use.

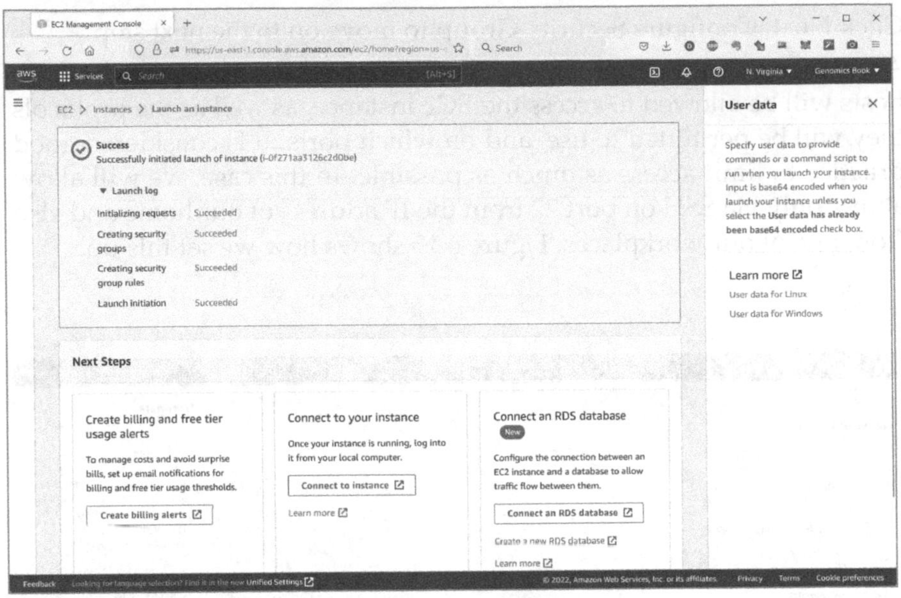

Figure 6.16: Reviewing the instance launch details

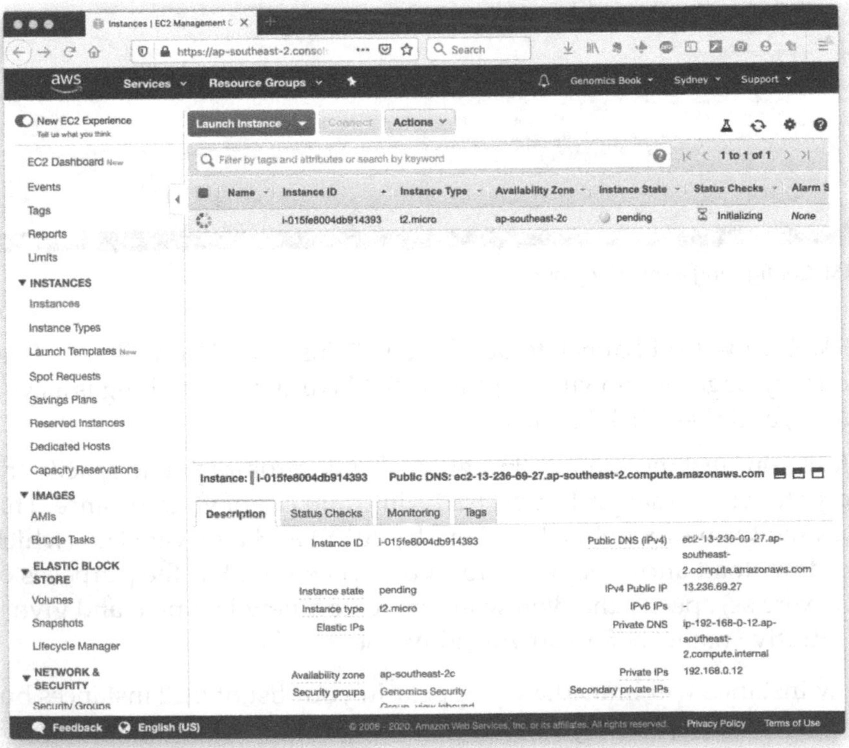

Figure 6.17: An EC2 instance starting up, showing in Pending state.

Shutting Down an Instance to Save Money

When you are not using an EC2 instance, you should shut it down so that you are not billed for running time. This is easy: simply go to the list of EC2 instances in the AWS Console, select the one you want, and choose Stop from the Actions menu (as shown in Figure 6.18). Do not choose Terminate since that will destroy the instance altogether and you would have to re-create it to use it again.

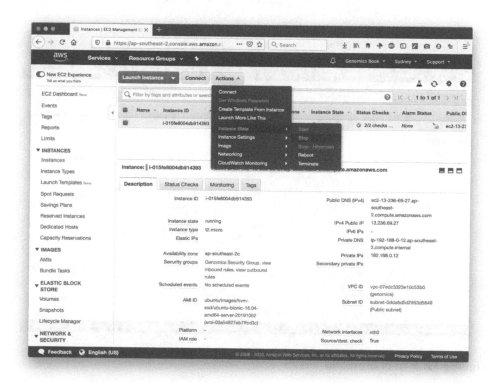

Figure 6.18: Stopping an instance to save running costs

When an instance is in the Stopped state, you will still pay for its storage volumes but not for the computing power of the instance itself.

Setting Up S3 Buckets

In almost any application, you'll need to set up some Simple Storage Service (S3) buckets for the purpose of storing input and output data. S3 is one of the oldest and most mature of the AWS services, and it's super easy to set up S3 buckets.

Here's the procedure:

1. Go to `console.aws.amazon.com`. Log in using your email address and password.

2. Click Services to see the entire list of services. S3 appears under the Storage section and is usually near the top, but you can also type it into the box at the top of the page to filter out everything else. Click the S3 link. You will see your bucket list, in a manner of speaking, as shown in Figure 6.19.

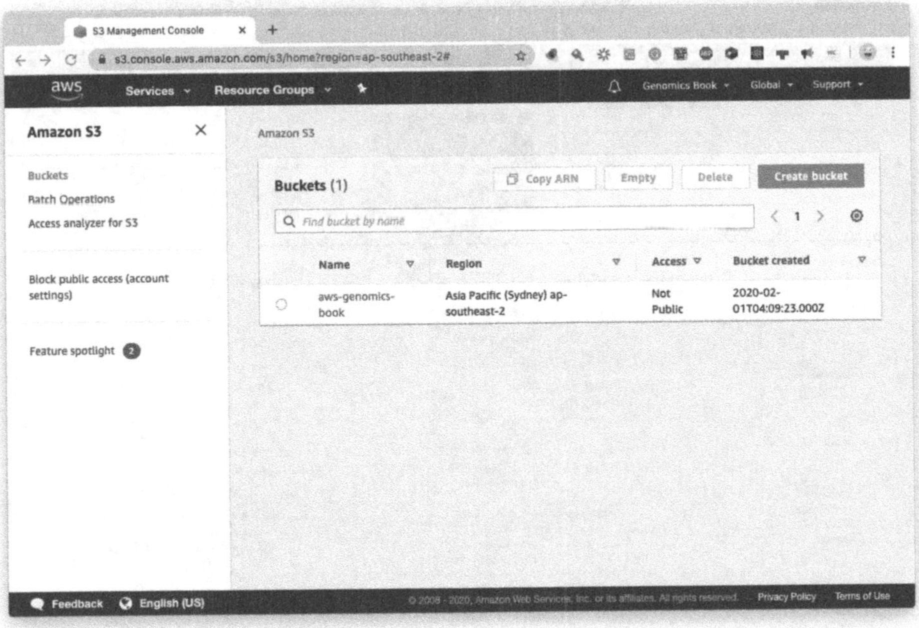

Figure 6.19: A list of all existing buckets

3. Give your bucket a name.

   The name has to be compatible with the Domain Name Service (DNS), which means it cannot end in a dash or a period. You have to use all lowercase letters, too. Interestingly, it also has to be unique across all bucket names in AWS S3, even those you don't control. For this reason, you might want to use a naming convention that helps satisfy this requirement. We name our buckets with the prefix *dw* or *cv*, for example. Figure 6.20 shows the naming of a new bucket.

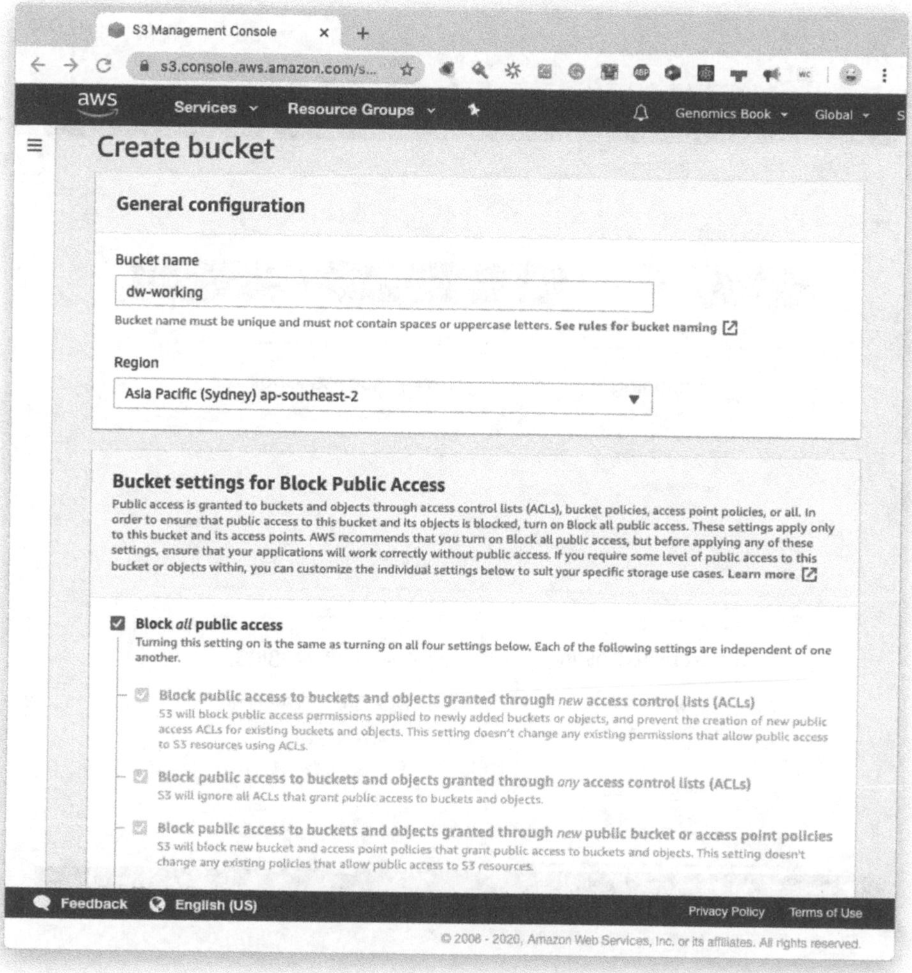

Figure 6.20: Naming a new bucket

4. Choose options for your bucket. The main one is versioning, which is turned off by default and probably should remain turned off for genomics work. If you turn it on, you'll be able to roll back to earlier versions of your files if you need to, but you'll also pay more for storage (and storage is already one of the big costs associated with genomics work on S3). Plus, versioning is not functionally appropriate, since your files will be either static (your input files) or able to be regenerated (your output files). Figure 6.21 shows the selection of options.

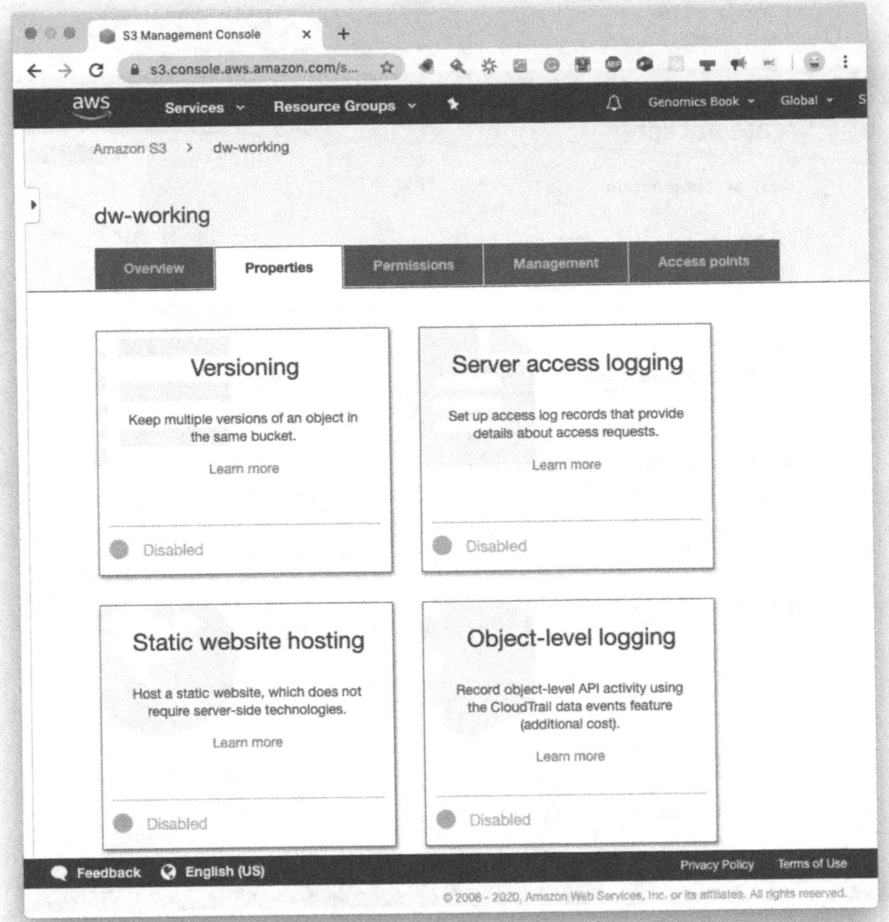

Figure 6.21: Choosing bucket options

5. Set the access permissions for the bucket. Generally, for private work, you shouldn't change anything on this page. You can grant access to specific people (or make individual files public, if you want to) later. Figure 6.22 shows the access-specification page.

6. Review the bucket settings and click the Create Bucket button to build the bucket as you've specified. Your bucket will appear in the list on the main S3 page. Figure 6.23 shows a newly expanded list of buckets.

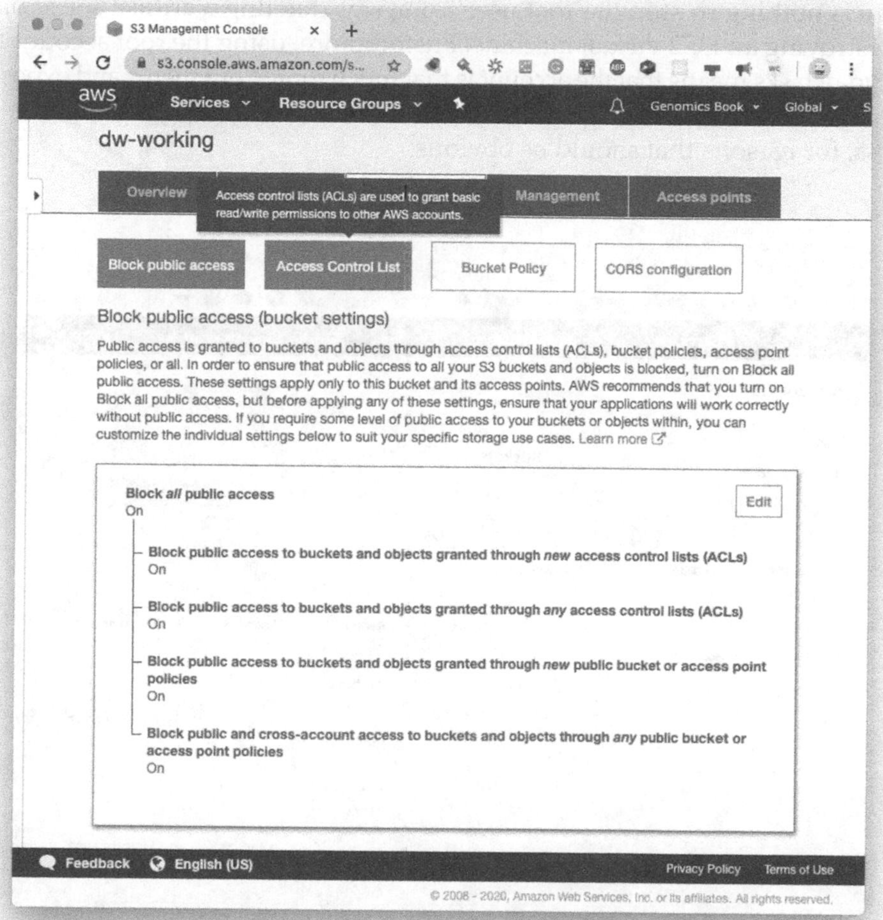

Figure 6.22: Declining the option to make the new bucket public

Configuring Your Account Securely

When you created your account, you established what is called the *root user*. The root user is essentially the ultimate administrator of your account. That user has access to absolutely everything and can create, remove, or alter any aspect of the account. While the so-called "god account" is super-handy and it may seem like the easiest way to get things done is to log in under that account, doing so is considered poor practice. For one thing, it is easy to make mistakes. There is nothing to stop the root user from, say, deleting a critical set of data or destroying an EC2 virtual machine. Furthermore, using the root account for everyday tasks means that the account is that much more "out there" and exposed to malicious hacking attempts. Black-hat hackers love getting administrative access, for reasons that should be obvious.

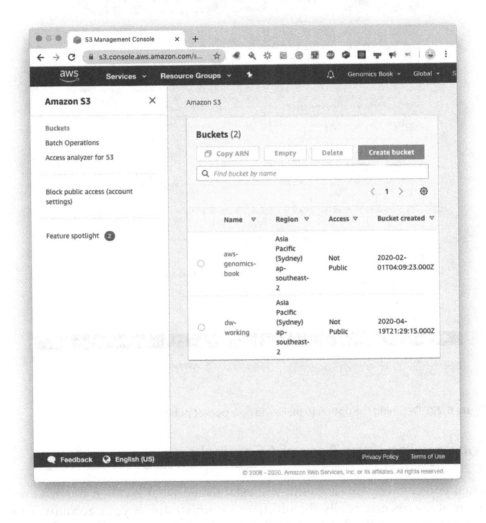

Figure 6.23: The new bucket in the list of all buckets

Therefore, good practice demands that we use the root account to create other accounts for each user of the system (both human and programmatic), each with a level of access and privilege appropriate to what they need to do—enough and no more.

To create secondary accounts, we need to use AWS Identity and Access Management (IAM). IAM is a service listed among the many others in the AWS Console. You can look for it in the list under Security, Identity, and Compliance, but it's probably quicker to search for it in the search box. Figure 6.24 shows the IAM main page.

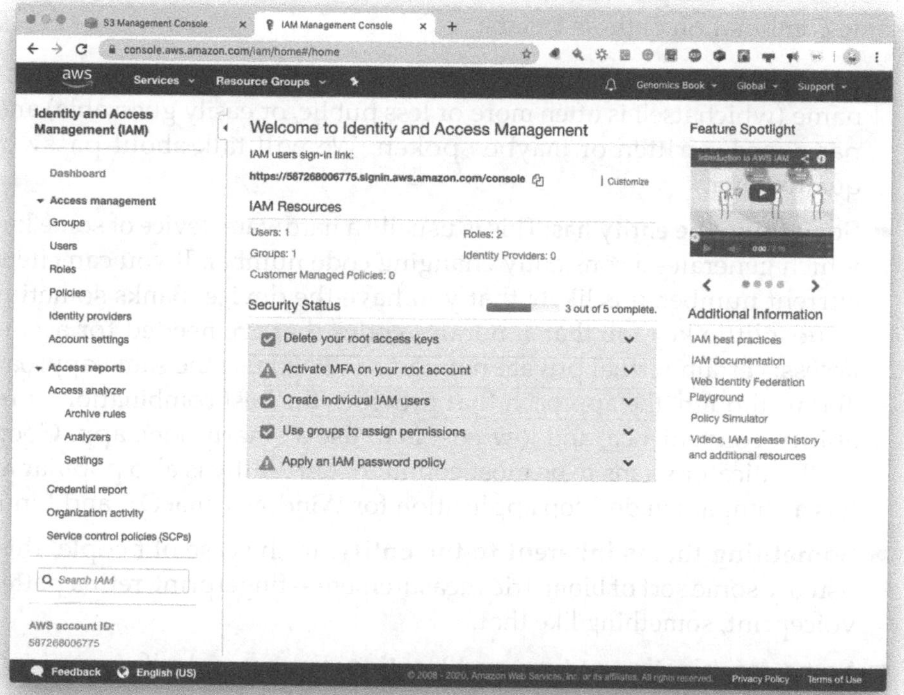

Figure 6.24: The main IAM page

Note that the IAM dashboard page is warning us about our security status: only three of the five recommended actions have been implemented. (These three steps are carried out automatically as part of account setup.) While we are here for the purpose of creating subsidiary users, we should take the opportunity to tighten our account's security by working through the remaining two steps: turning on multifactor authentication and establishing an AWS IAM password policy.

Turning On Multifactor Authentication

This is not a security textbook, but it is worth taking a look at why setting up AWS MFA is a good idea. If nothing else, information security is an interesting topic and we are curious people.

Authentication confirms that someone (or something, in the case of a piece of software) is who they say they are. If a system can determine, with a high degree of confidence, that an assertion about identity ("I am David Wall") is true, then it can move on to the problem of access, which is to say letting the user see and do things appropriate to their identity.

A system establishes the authenticity of an identity claim (performs authentication) by evaluating characteristics of the entity making it. These boil down to some combination of these factors:

- **Something the entity knows**: This is usually the combination of a username (which itself is often more or less public, or easily guessable) and a password, written or maybe spoken. We will talk about password quality later.

- **Something the entity has**: This is usually a hardware device of some kind, which generates a constantly changing code number. If you can cite the current number, it is likely that you have the device. Banks sometimes issue a little key fob that generates codes that are needed for account access; certain virtual private networks (VPNs) use the same approach. Today, though, the approach that provides the best combination of reliability, convenience, and low cost is to use a smartphone app. Google Authenticator seems to be most common, but Authy is also popular and has a companion desktop application for Windows, macOS, and Linux.

- **Something that is inherent to the entity**: In the case of people, this is usually some sort of biometric measurement—fingerprint, retina pattern, voiceprint, something like that.

- **Some combination of time and location**: For example, if a user is supposed to be accessing something from a particular corporate network during certain business hours, their claimed identity might be challenged if they appear to be coming in from across the Internet or at a time outside the access window.

In the case of our AWS account, we will set up the first two of those factors. We already have a username and password in place on our root account, and we will supplement that with hardware authentication based on Google Authenticator running on a smartphone. This will be two-factor authentication (2FA), the simplest form of MFA, and it's good enough for almost everything, including the protection of the personal genome information we are working with here.

To set up MFA on your AWS account, open the Activate MFA On Your Root Account section of the Security Status list and click the Manage MFA button. You will be taken to the Your Security Credentials page, where you should click the Activate MFA button (Figure 6.25).

Choose Virtual MFA Device from the Manage MFA Device pop-up, and click Continue (Figure 6.26).

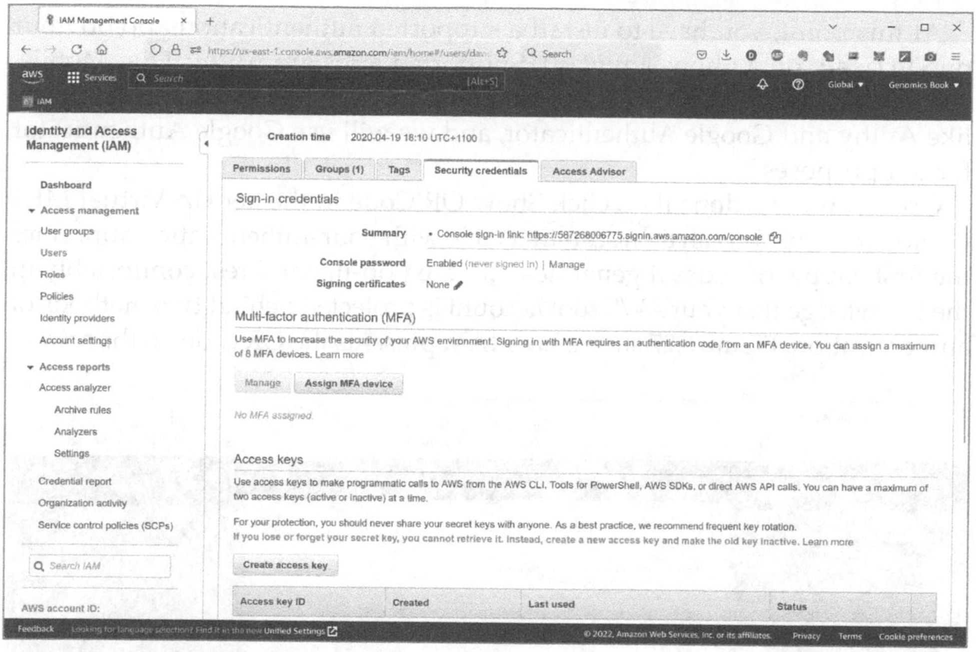

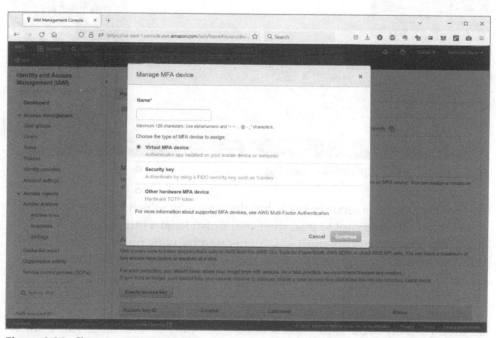

Figure 6.25: Activating multifactor authentication

Figure 6.26: Choosing an MFS device

At this point, you have to install a supported authentication app on your phone (well, on a phone anyway). Supported apps are Authy, Duo Mobile, LastPass Authenticator, Microsoft Authenticator, and Google Authenticator. We like Authy and Google Authenticator, and we will use Google Authenticator for our purposes here.

Once you have done that, click Show QR Code in the Set Up Virtual MFA Device pop-up. You can then scan that code with your authentication app, enter the first couple of codes it generates into the pop-up, and rest comfortably in the knowledge that your AWS root account is protected behind two methods of authentication. Figure 6.27 shows how IAM presents the QR code on the screen.

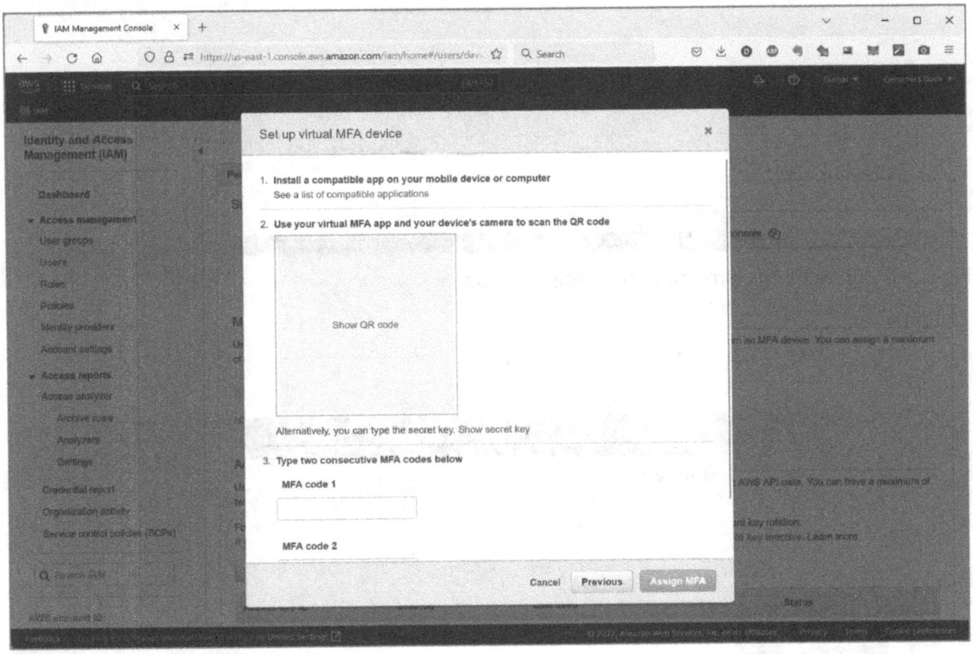

Figure 6.27: Getting a QR code for scanning

Again, this is not a security book, but the means by which this kind of authentication works is too cool to pass over without at least a little exploration. Consider that you can use six-digit keys generated by your phone, even when the phone is disconnected from the Internet—even when it is in flight mode or in a Faraday cage. How can that be?

Authentication tools implement algorithms called HOTP and/or TOTP (Google Authenticator implements both). These algorithms apply a cryptographic hash method (usually SHA-1) to a secret key known only to both sides (the authentication app and AWS IAM). It is this secret key that is communicated to your authentication app when you scan the QR code in IAM. The HOTP/TOTP algorithm, using the secret key, generates the same series of six-digit keys when it is run on AWS and when it is run on the app. The fourth code generated by the app is the same as the fourth code generated by the implementation of HOTP/TOTP on AWS; the same is true for the fifteenth and three-hundredth run on each side. If AWS knows that a given authentication attempt for a certain user is the 564th, it knows what code the algorithm generates for that iteration and therefore what the user should enter. The only remaining problem is keeping the counts in sync so that AWS knows what run of the algorithm should be valid. TOTP—which stands for "time-based one-time password"—solves this problem by assuming synchronized (or nearly so) clocks both on the phone running the app and in the AWS IAM environment. Network Time Protocol (NTP) makes this easier to achieve than it used to be. The other algorithm, HOTP (its seriously awkward expansion is HMAC-based one-time password, where HMAC is Hash-based Message Authentication Code) does not assume a common clock and requires synchronization of the "which run is this" number. It also allows some leeway in the run count, up or down.

Establishing an AWS IAM Password Policy

The next step in tightening up the security situation on your AWS environment is to establish a password policy to ensure that users' passwords are obscure enough to resist guessing and brute-force attacks and are changed according to a schedule. Figure 6.28 shows IAM reminding you to set up a policy,

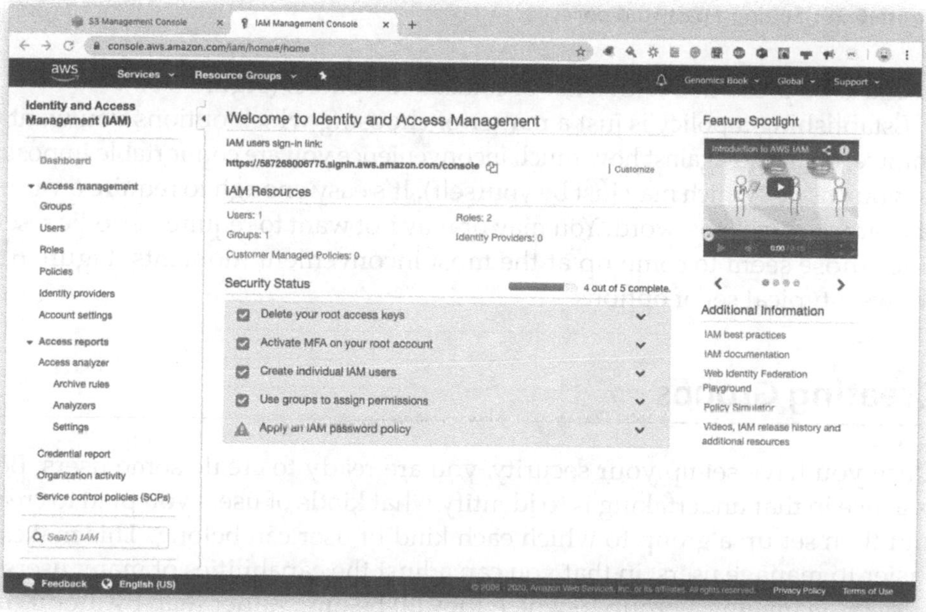

Figure 6.28: IAM wants you to set up a password policy for enhanced security.

To set up a password policy on your AWS account, open the Apply An IAM Password Policy section of the Security Status list and click the Manage Password Policy button. You will be taken to the Account Settings page, shown in Figure 6.29, where you should click the Set Password Policy button.

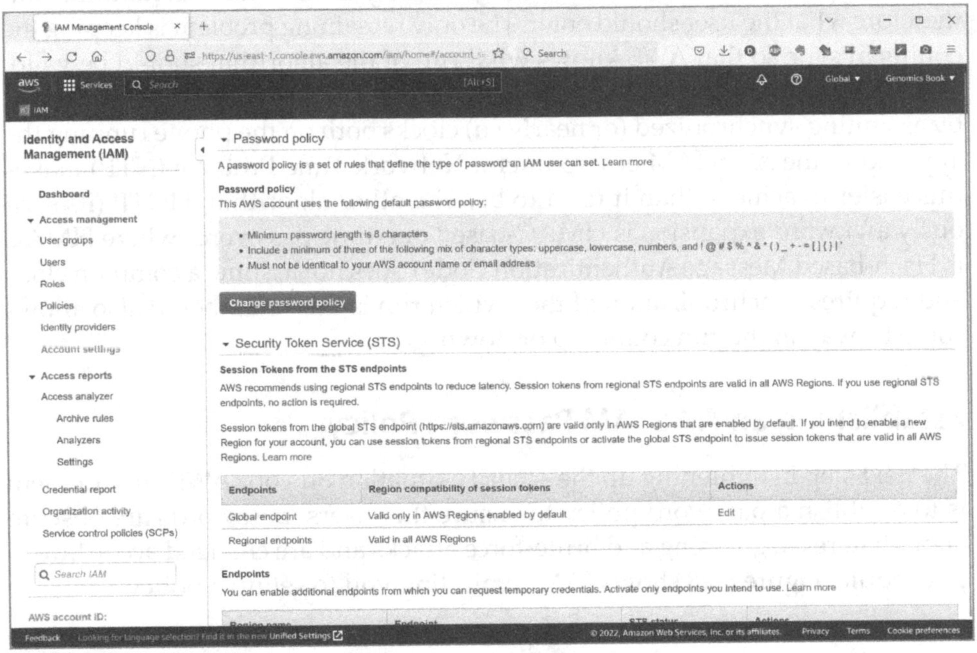

Figure 6.29: Setting a password policy

You'll be presented with a list of selectable options (Figure 6.30).

Establishing a policy is just a matter of choosing which options you want to enforce, balanced against how much inconvenience you are comfortable imposing on your users (which may just be yourself). It's easy enough to require a variety of characters in a password. You may or may not want to require periodic resets, since those seem to come up at the most inconvenient moments. Figure 6.31 shows a typical set of options.

Creating Groups

Once you have set up your security, you are ready to create some users. Best practice in that undertaking is to identify what kinds of users you plan to create and then set up a group to which each kind of user can belong. This makes it easier to manage users, in that you can adjust the capabilities of many users at once by managing a group to which they all belong, rather than by modifying the characteristics of the users individually.

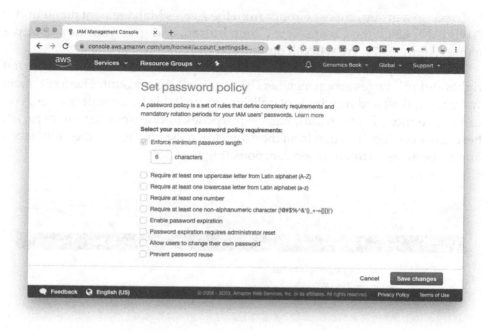

Figure 6.30: Specifying characteristics of the password policy

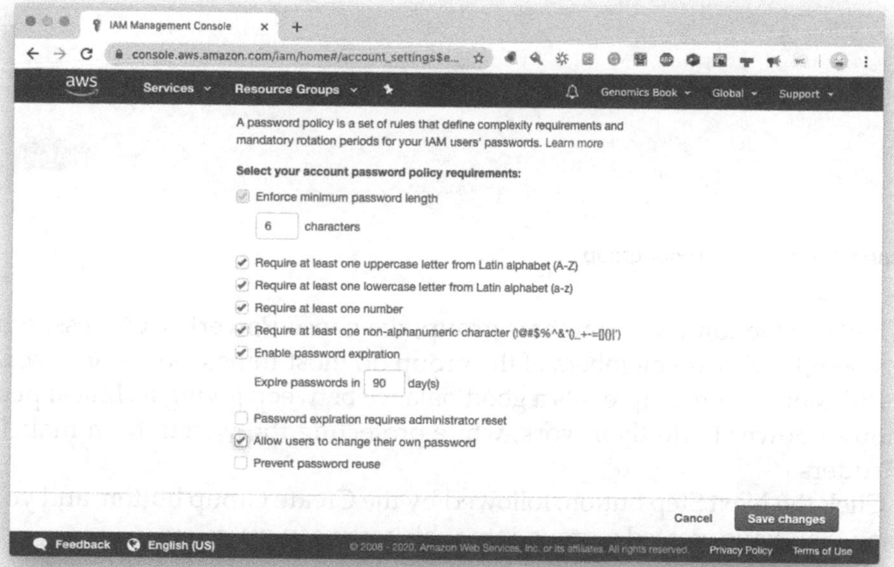

Figure 6.31: A selection of password policy characteristics

To set up a group, choose Groups from the Access Management menu under AWS IAM. Click the Create New Group button, and give your new group a name (Figure 6.32). We will call our group GenomicsResearchers.

The next step is to attach policies to the group, which is how we define what rights and privileges group members have in the AWS account. There are many ways to do this, and hundreds of policies are available by default (as well, you can create more if you want). For the purposes of our undertaking here, it's best to choose Job Function from the drop-down list box next to the Filter label. You'll see an assortment of job functions (Figure 6.33).

Figure 6.32: Naming your group

For our GenomicsResearchers group, we chose PowerUserAccess, which lets people who are members of the group do most things—but not alter IAM permissions, for example. It's a good balance between giving technical people enough power to do their work, while protecting the system from malicious intruders.

Click the Next Step button, followed by the Create Group button, and you're done. You have created a group into which you can put users.

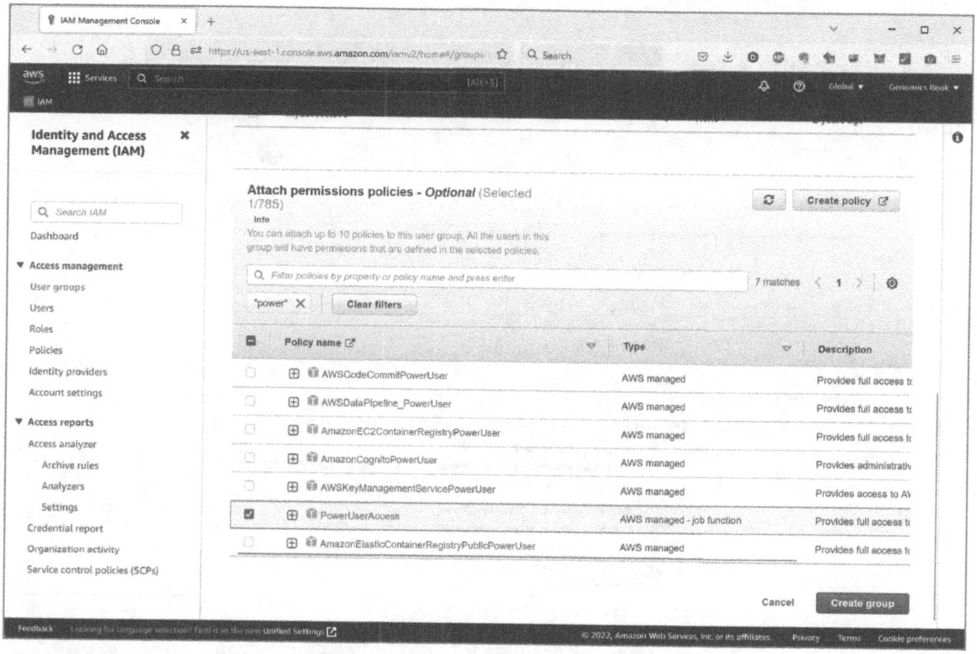

Figure 6.33: Assigning the standard PowerUserAccess policy to the new GenomicsResearchers group

Creating Users

Creating users is not hard. All you need to do is choose Users from the Access Management menu under AWS IAM. Click the Add User button, and give your new user a name. You can choose to create multiple users at once, which is what we did (Figure 6.34).

Click the Next: Permissions button to move on to the next step, where you can add your new users to a group (Figure 6.35). We added our new users to the GenomicsResearchers group that we created previously.

Subsequent steps allow you to add tags to your agents (not necessary in our case) and review what you're about to do before you create the new users.

Once AWS IAM has created the users you specified, it will show you a Success page that shows important characteristics of the users (Figure 6.36). It will display the access key ID, secret access key, and initial password. The username and password are obviously important for logging into the console, but you need the access key ID and secret access key to do anything programmatically. The Success display in the user creation process is the only time it is possible to display those values, so copy them somewhere safe or download and save them in CSV form.

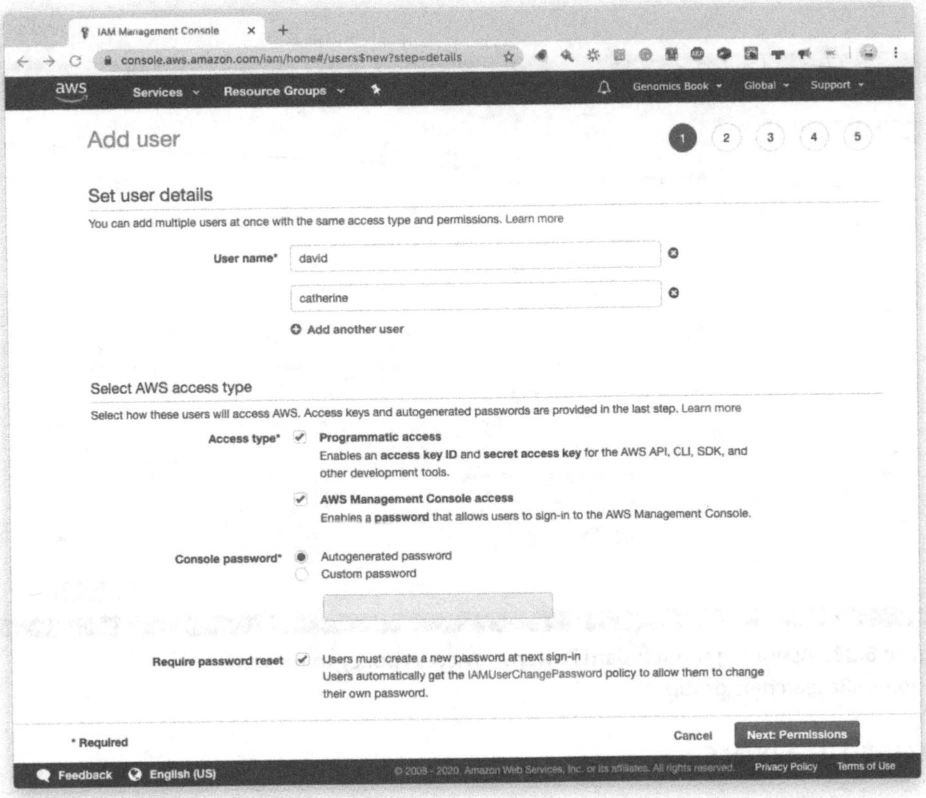

Figure 6.34: Creating two new users

Setting Up Your Client Environment

Now that you have set up your AWS environment securely, you are able to connect to it from whatever computer you use locally—the notebook or desktop computer that sits on your working desk. Setting up these connections will enable you to use both your local tools (for writing and that kind of thing) and the remote resources in the AWS cloud for storage and manipulation of large data sets. If you do it right, the integration between your local environment and the remote one will be seamless.

Connecting to an EC2 Instance

Connecting to your newly created EC2 instance is easy. AWS even provides explicit instructions: just click the Connect button near the top of the EC2 instances list.

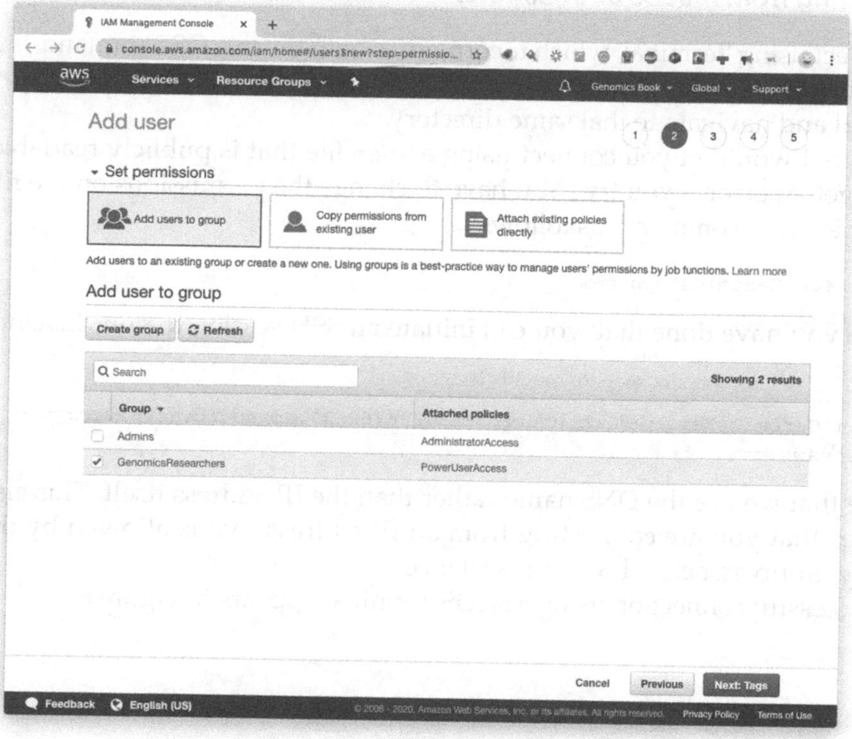

Figure 6.35: Assigning the new users to a group

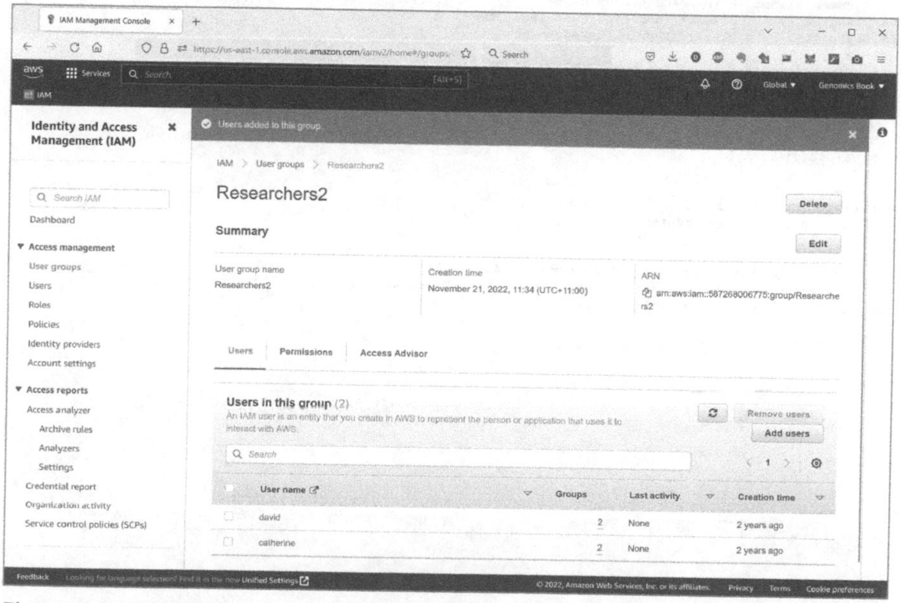

Figure 6.36: Two new users, successfully created

Connecting from macOS or Unix/Linux

To connect using Terminal, which incorporates both the macOS command-line interface and its SSH connectivity, put your .pem file in a folder. Then, open Terminal and navigate to that same directory.

Terminal won't let you connect using a .pem file that is publicly readable. (You'll get an error if you try.) You have to change the permissions on the file using the chmod command, as follows:

```
chmod 400 GenomicsKeys.pem
```

Once you have done that, you can initiate an SSH session to your instance like this:

```
ssh -i GenomicsKeys.pem ubuntu@ec2-13-236-69-27.ap-southeast-2.compute
.amazonaws.com
```

Note that we use the DNS name rather than the IP address itself. This also assumes that you are connecting from an IP address that is allowed by the security group associated with the instance.

A successful connection using MacOS Terminal appears in Figure 6.37.

```
● ● ● ⌂ davidwall — ubuntu@ip-192-168-0-12: ~ — ssh -i GenomicsKeys.pem ubuntu@ec2-13-236-69-2...

192-168-1-4:~ davidwall$ chmod 400 GenomicsKeys.pem
192-168-1-4:~ davidwall$ ssh -i GenomicsKeys.pem  ubuntu@ec2-13-236-69-27.ap-southeast-2.compute
.amazonaws.com
Welcome to Ubuntu 18.04.3 LTS (GNU/Linux 4.15.0-1051-aws x86_64)

 * Documentation:  https://help.ubuntu.com
 * Management:     https://landscape.canonical.com
 * Support:        https://ubuntu.com/advantage

  System information as of Sun Feb  2 03:44:01 UTC 2020

  System load:  0.0              Processes:            86
  Usage of /:   11.0% of 9.63GB  Users logged in:      0
  Memory usage: 14%              IP address for eth0:  192.168.0.12
  Swap usage:   0%

0 packages can be updated.
0 updates are security updates.

Last login: Sun Feb  2 03:23:12 2020 from 123.243.26.138
To run a command as administrator (user "root"), use "sudo <command>".
See "man sudo_root" for details.

ubuntu@ip-192-168-0-12:~$ ▉
```

Figure 6.37: Connecting to an EC2 instance via MacOS Terminal

Connecting from Windows

When you are working under Microsoft Windows, it is best to use an application called PuTTY rather than the SSH client that is built into the command line. PuTTY (the capitalized TTY refers to "teletypewriter," a very old standard for text interfaces) is the standard tool for SSH work. You can download PuTTY free of charge from `putty.org`.

Before you can connect to your EC2 instance, you must convert the `.pem` file you downloaded from AWS to a form PuTTY can use. PuTTY comes with a tool called PuTTY Key Generator (or PuTTYGen) that will convert a `.pem` file into a PuTTY-compatible `.ppk` file.

Load your `.pem` file into PuTTY Key Generator, as shown in Figure 6.38. Then, click the Save Private Key button and specify a new file (with a `.ppk` extension).

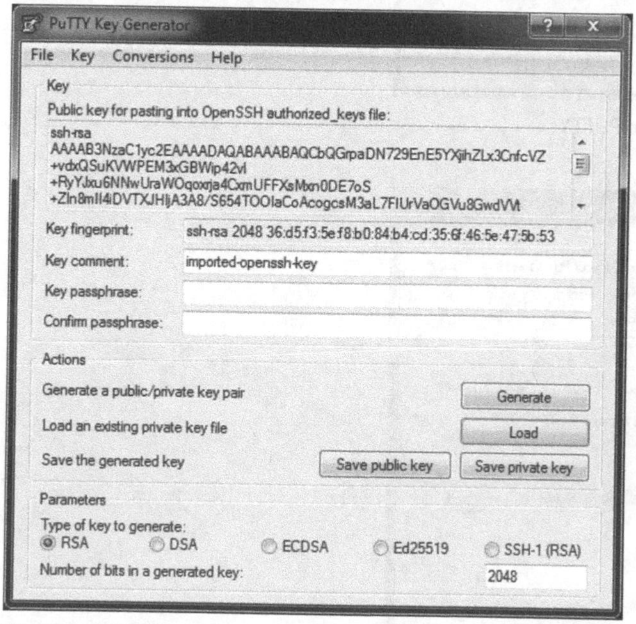

Figure 6.38: Creating a `.ppk` file with PuTTY Key Generator

In PuTTY itself, navigate to Connection, SSH, Auth in the tree. Use the Browse button next to the Private Key File for Authentication box to specify the `.ppk` file you just created. Figure 6.39 shows what this looks like.

On the Session page of PuTTY, put the DNS name of your EC2 instance in the Host Name box, as shown in Figure 6.40.

Click Open, and you should get in, as shown in Figure 6.41.

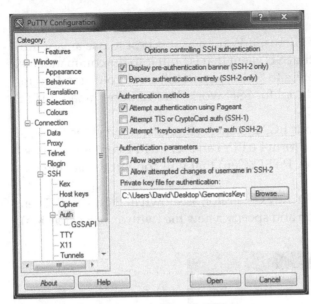

Figure 6.39: Adding a .ppk file to PuTTY

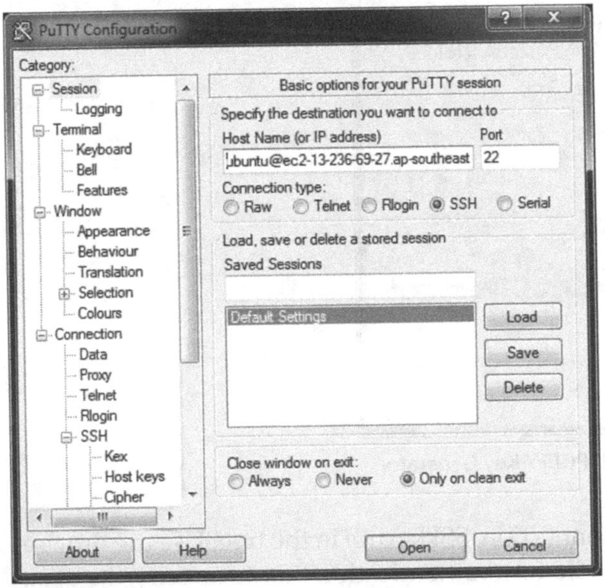

Figure 6.40: Specifying an EC2 instance in PuTTY

Making S3 Buckets Available Locally

While it is possible to use the S3 console to access S3 buckets and their contents, you will probably find it easier to work with S3 storage if you mount buckets locally as a storage volume so they appear like any other storage location.

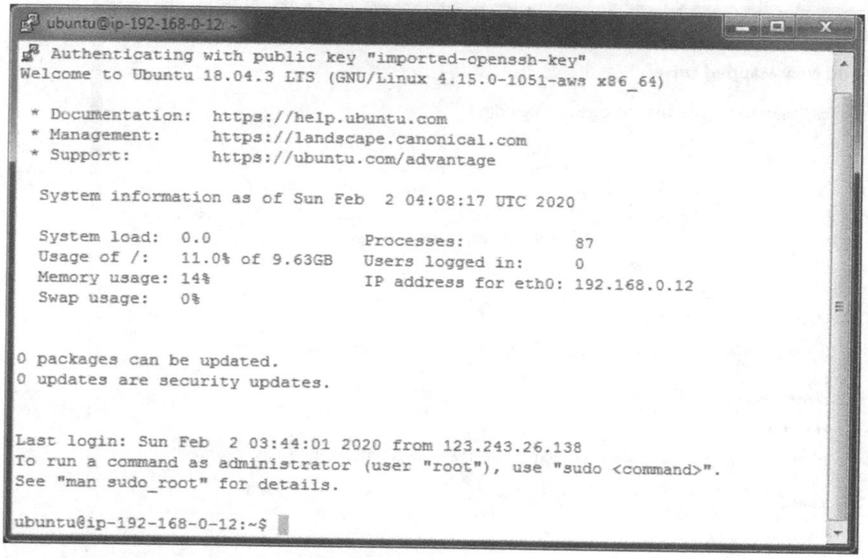

```
ubuntu@ip-192-168-0-12: ~                               _  □  X
 Authenticating with public key "imported-openssh-key"
Welcome to Ubuntu 18.04.3 LTS (GNU/Linux 4.15.0-1051-aws x86_64)

 * Documentation:  https://help.ubuntu.com
 * Management:     https://landscape.canonical.com
 * Support:        https://ubuntu.com/advantage

  System information as of Sun Feb  2 04:08:17 UTC 2020

  System load:  0.0                Processes:            87
  Usage of /:   11.0% of 9.63GB    Users logged in:       0
  Memory usage: 14%                IP address for eth0: 192.168.0.12
  Swap usage:   0%


0 packages can be updated.
0 updates are security updates.


Last login: Sun Feb  2 03:44:01 2020 from 123.243.26.138
To run a command as administrator (user "root"), use "sudo <command>".
See "man sudo_root" for details.

ubuntu@ip-192-168-0-12:~$
```

Figure 6.41: A successful PuTTY connection

Mounting an S3 Bucket as a Windows Drive

Our preferred tool for working with S3 buckets under Windows is TNTDrive. It's easy to set up and has been completely reliable.

Setting up TNTDrive is simple. Download the installer executable from tntdrive.com and run it. You have to reboot after installation completes.

Once you have rebooted, you can run TNTDrive and give it your access key ID and secret access key. With those, TNTDrive will connect to S3 and show your bucket hierarchy. You can then click Add New Mapped Drive in TNTDrive to map a drive letter to the top of your S3 environment (so you can access the entire folder tree) or to some subordinate point. This is shown in Figure 6.42.

Mounting an S3 Bucket Under macOS and Linux

The options for mounting an S3 bucket as a drive under macOS are generally not as slick as what is available for Windows. There is, however, a perfectly reliable piece of software called S3FS-FUSE available on GitHub. You can download and install it and then access your S3 environment from within the macOS Finder, same as you would with any network drive. There's no special client, which is maybe even better than the TNTDrive approach.

While there are a few ways to install S3FS-FUSE under macOS (you can download and build the source manually, if you want), the easiest way is with Homebrew, as shown here:

```
brew cask install osxfuse
```

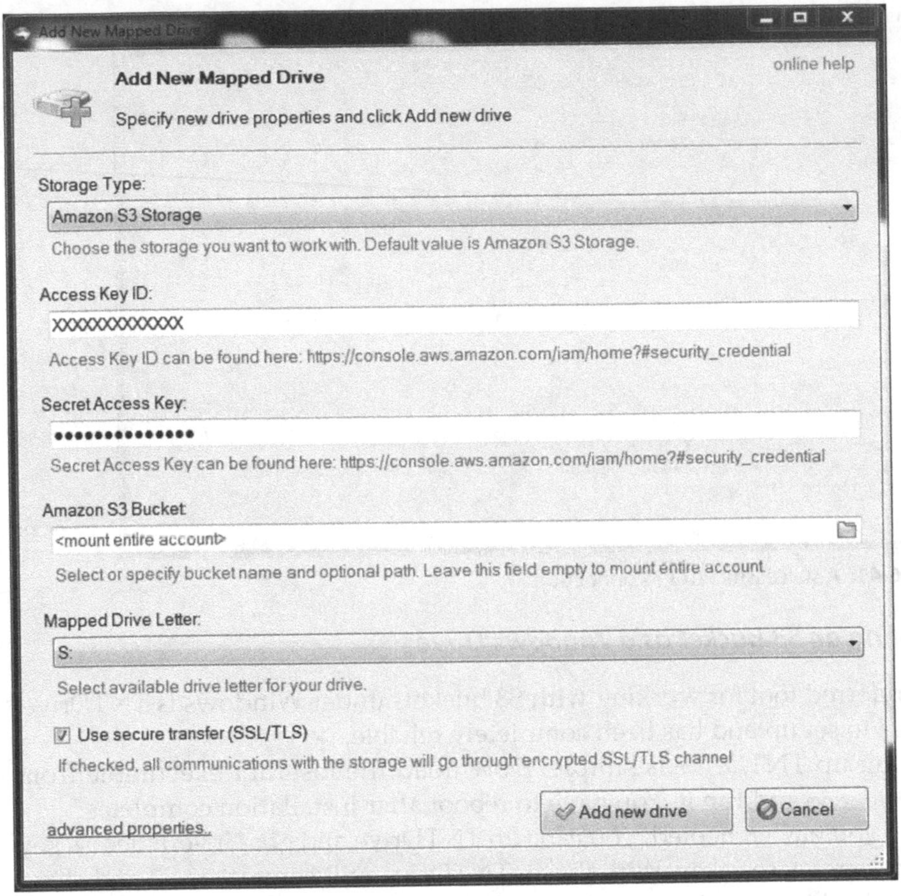

Figure 6.42: Attaching an S3 location to Windows

At this point, you must reboot your Mac, as follows:

```
brew install s3fs
```

Then, you need to make S3FS aware of your AWS access key ID and secret access key. Again, there are a couple of ways, but the easy way is as follows:

```
echo <Your access key id>:<Your secret access key> <Path to your home
directory>/.passwd-s3fs
chmod 600 <Path to your home directory>/.passwd-s3fs
```

The method for setting up S3FS under the various Linux flavors is similar. With that, you should be able to see and browse folders in Finder, same as any other local or remote drive. The main difference you will notice between local and S3 volumes will be the file transfer latency.

Summary

In this chapter, you set up an AWS account and configured storage and computing resources in it. You learned how to configure those resources for maximum security and how to connect to them from a client machine so you can do real work using both local and AWS tools.

The next chapter describes the Linux environment in which we will do most of our genome analysis work.

Linux and AWS Command-Line Basics for Genomics

Genomics involves manipulating large amounts of data. The Linux operating system is particularly suited to manipulating large files and has become the operating system of choice for manipulating and extracting information from all types of biological data, including genomics, proteomics, and other "omics" data. The Linux command-line interface allows easily visualization and manipulation of very large text files of millions of lines, whereas it is difficult to perform similar tasks using Windows Excel or other Windows tools. In addition, a lot of bioinformatics software is available only on Linux. Therefore, the first skill required to analyze genomics data is to become proficient with Linux. By the end of this chapter, you will be familiar and comfortable with all key Linux concepts necessary to run your genomics analyses.

As well, you can install an AWS package that enables you to manage your AWS resources directly from the command line. With the AWS CLI Tools installed, you can do such things as copy files from your local filesystem to an S3 bucket, start and stop EC2 instances, and manage user access privileges.

Selecting a Linux Distribution

You are probably aware that Linux comes in many flavors, called *Linux distributions,* such as Ubuntu, Red Hat, etc. To understand this concept, let's start with defining what an operating system is.

An *operating system* is a program, or software, that interfaces with all basic hardware components of a computer and provides tools that enable you to use your computer. It is composed of the following:

- **A kernel:** The kernel is the core part of the operating system, which interfaces directly with the central processing unit (CPU), memory, hard disks, network adapters, and other hardware devices of your computer.

  Examples of kernels are the Windows kernel, the macOS kernel, and the Linux kernel. The Linux kernel is free and open source and was first developed in 1991 by a Finnish student named Linus Torvalds. The Linux kernel shares many similarities with the Unix operating system, developed by Bell Labs at AT&T in 1969, and is available for free under the GNU (standing for "GNU's Not Unix") General Public License.

  Most users do not interact with the kernel itself but with intermediary programs that call kernel functions, such as the intermediary programs listed next.

- **A command-line interface:** This interface, also called the *text-mode shell*, enables you to type in commands to list, copy, or rename files; to launch programs; etc. This command-line interface can be accessed in Linux or macOS by launching the program "terminal" or in Windows by typing cmd in your Windows taskbar. With Linux, several Linux command-line interfaces are available, such as the Bourne shell (sh), the Bourne Again shell (Bash), the C shell (csh), and the Korn shell (ksh).

- **A graphical user interface (GUI):** The GUI allows you to perform actions by manipulating icons and menus and using the functions of a mouse. Windows and macOS shine with their very extensive GUI facilities. Linux relies on a primitive GUI called X Windows System (named X). To extend this primitive GUI, several Linux desktop environments have been developed that provide a better overall Linux GUI environment, such as the K Desktop Environment (KDE) or the GNU Object Model Environment (GNOME).

- **Utilities:** These include text editors, file managers, programs to clean up or defragment disks, utilities to install or deinstall programs, etc.

- **Libraries:** Linux libraries comprise programming functions that are used by various programs.

How does Linux differ from Windows and macOS? Linux, as already mentioned, is free and open source (meaning the source code of Linux is freely available), whereas Windows and macOS are commercial, proprietary operating systems. macOS, like Linux, is Unix-based, and therefore the command-line interfaces of Linux and macOS accept many of the same commands. However, the macOS GUI does not use the X Windows System, so macOS applications cannot run on Linux.

Windows and macOS have an emphasis on the graphical user interface, while Linux has a historical focus on the command-line interface. Linux tends to be less resource intensive and therefore is the operating system of choice on servers. Although several GUI interfaces are available on Linux, the use of Linux on desktops or laptops is sometimes impeded by the lack of availability of drivers for the latest peripherals or the fact that popular applications such as Microsoft Office or Adobe are not available on Linux. Linux alternatives exist, such as OpenOffice.org or LibreOffice, but are less popular.

The Linux operating system is available under different distributions: Ubuntu, Red Hat, CentOS, openSUSE, and others. What is a Linux distribution? Each Linux distribution contains the Linux kernel (always the same kernel, albeit some distributions might use different releases or patches of the Linux kernel), as well a combination of utilities and software. So, in addition to the Linux kernel, a Linux distribution typically includes the core Unix tools, such as the GNU tools and the X Windows System, additional software, startup scripts, documentation, a desktop environment, and a package management system.

It is important to make yourself familiar with the package management system. With most Linux distributions, software is made available as groups of files called *packages*. The package management system maintains a local database of installed files and includes a set of tools that facilitates the installation, upgrade, configuration, and removal of software. There are two main Linux package manager systems:

- **The RPM Package Manager (RPM):** The RPM is used by many Linux distributions, such as CentOS, openSUSE, SUSE Enterprise, and Red Hat Enterprise.

- **The Debian package manager:** The Debian package manager is mainly utilized by the Debian, Ubuntu, and Ubuntu-derived Linux distributions.

Finally, some Linux distributions use their own package management methods, such as the Slackware Linux distribution that uses packages called *tarballs* that are created by the standard tar utility.

Table 7.1 lists the main Linux server distributions. Server distributions do not provide a desktop environment and are therefore very suitable for running on the cloud.

Table 7.1: Main Linux Server Distributions

DISTRIBUTION	AVAILABILITY	PACKAGE FORMAT
Debian	Free	Debian
Ubuntu	Free	Debian
CentOS	Free	RPM
Red Hat Enterprise	Business	RPM
SUSE Enterprise	Business	RPM
openSUSE	Free	RPM

Debian is one of the oldest distributions and is used by lots of educational and government organizations. Ubuntu is a popular distribution derived from Debian and has both a server and a desktop version. Red Hat and SUSE Enterprise are open-source distributions developed for commercial use. The German company SUSE sponsors openSUSE and commercially supports SUSE Enterprise. The CentOS distribution is free and derived from Red Hat. Stability is important for servers, as they often run the production environments of businesses. Most distributions in Table 7.1 issue new releases approximately every 2 years, except openSUSE, which has new releases every 8 months, and Ubuntu, every 6 months. To cater for enterprise-type servers, Ubuntu also issues Ubuntu LTS editions every two years, where LTS stands for "Long Term Support."

Accessing Your AWS Linux Instance from Your Local Computer

This section explains how to access a remote AWS instance from your personal computer.

From Windows

Several programs allow you to open a window that gives you access to the command-line interface of your AWS Linux instance and therefore act as a terminal emulator of your instance. The most common free program is PuTTY, which can be downloaded from its website. It performs a secure remote login (called SSH, as Secure Socket Shell) to a remote computer, using the Telnet protocol. Note that Windows 10 and 11 now contain some of these functionalities that can be activated through a set of menus, but many people prefer PuTTY rather than digging through many menus. Launching PuTTY opens the configuration window, as shown in Figure 7.1. Click Session on the left, enter the IP address of your AWS instance, enter **22** for the port (this is a requirement for the SSH login), select SSH, and click the Open button. This will open a terminal window, and you will be asked for your password. You will then be connected to your AWS instance. Note that the first time you connect to a remote computer or instance, you receive a warning stating that you have not connected to this computer before. Click Accept to establish the connection.

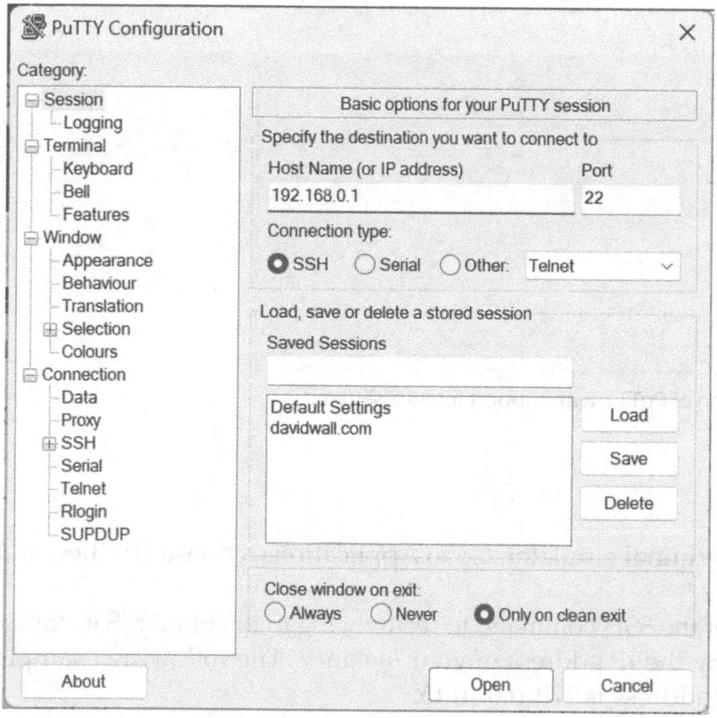

Figure 7.1: Configuration window of PuTTY

If you want to go beyond the instance command line and launch a graphical application, you will need to install an X server for Windows, such as Xming. Download (`sourceforge.net/projects/xming`) and install Xming with all the default settings. Then modify the PuTTY configuration as shown in Figure 7.2: expand the SSH category on the left, select X11, and select the option Enable X11 Forwarding. When you click the Open button, an X terminal window will open. You can use this window as a normal terminal window for usual Linux command lines—the only difference is that it will be slower and display some information in color—or use it to launch graphical Linux applications such as graphical text editors. Note that not all graphical Linux applications are supported and that the execution can be quite slow.

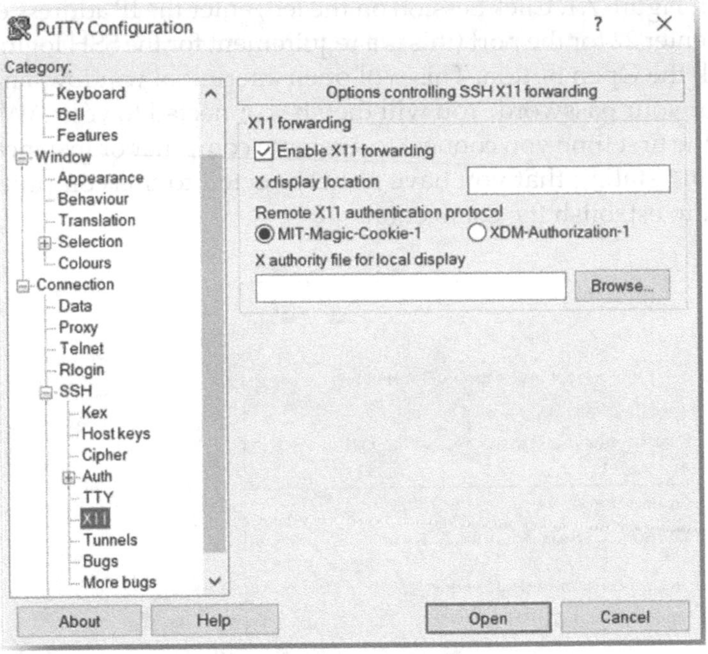

Figure 7.2: Configuration of PuTTY with support for an X server

From macOS

macOS includes a terminal emulator. Go to Applications, choose Utilities, and then open Terminal.

You can then enter the SSH command to securely log in to your AWS instance: type **ssh** followed by the IP address of your instance. The following example assumes that the IP address is 170.160.10.15:

```
ssh 170.160.10.15
```

Other free software extends the capabilities of the macOS embedded terminal emulator, such as iTerm2, ZOC, vSSH Lite, and OpenSSH.

Options for Setting Up Linux on Your Personal Computer

As mentioned, it is completely possible to connect to your AWS Linux instance from a non-Linux desktop, such as Windows or macOS. The support of the X server for graphical applications, however, can be tricky, as many Linux applications support it in a patchy way. In addition, the execution of the applications can be frustratingly slow. In the past, the options for running Linux on a desktop or laptop were either to fully migrate to Linux or to have a dual-boot, which is not very flexible. Nowadays, with the advent of virtualization software such as VMware that allows you to create virtual machines, it is possible to set up a full Linux environment running on your Windows or macOS desktop in a matter of minutes. This allows you to practice Linux locally or run some Linux graphics-intensive applications in a faster way.

The Linux distributions most suitable to desktops have a big focus on the GUI with an excellent desktop environment and are most suitable to beginners. Ubuntu provides a popular desktop Linux distribution that has a lot of online support from the community. Ubuntu ships with the Gnome desktop environment but accepts other desktop environments as well. Linux Mint and elementary OS are derived from Ubuntu and are also popular; Linux Mint has a Windows look and feel, while the beautiful interface of elementary OS is closer to the macOS user interface. A new distribution based on Debian, MX Linux, is gaining in popularity, as it has a relatively small footprint, which means it can run on older laptops and is easy to use and reliable. Finally, passionate Linux users come to Linux to be different and not follow the Windows and macOS crowds. For these users, Ubuntu or Linux Mint are definitely too mainstream, and they prefer one of the many niche desktop Linux distributions.

Figure 7.3 displays an Ubuntu virtual machine running, thanks to the virtualization software VMware, on a Windows 10 laptop. VMware is free for noncommercial environments.

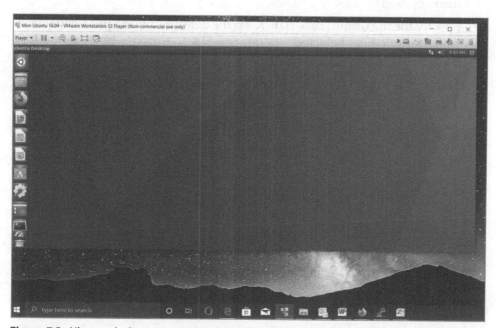

Figure 7.3: Ubuntu desktop running as a virtual machine (with VMware) on Windows 10

The steps for the installation and configuration are as follows:

1. Enable VT/VTd in Windows BIOS, as follows:

   a. Go to the Settings menu and select Update & Security.

   b. Select Recovery.

   c. Under Advanced Setup, select Restart Now.

   d. Select Troubleshoot.

   e. Select Advanced Options.

   f. Select UEFI Firmware Settings

   g. Under Restart to Change UEFI Firmware Settings, select Restart ➤ Enable Options VT/VTd.

2. Download VMware Workstation Player for Windows 64-bit operating systems.

3. Download `Ubuntu-xx.xx.x-desktop-amd64.iso`.

4. Install VMware by executing the downloaded VMware file and selecting Use VMware Workstation XX Player For Free For Non-commercial Use.

5. Launch VMware and select Create A New Virtual Machine.

6. Select Installer Disc Image File and browse to point to `ubuntu-xx.xx.x-desktop-amd64.iso`.

7. Specify the maximum disk size. Keep the default option Split Virtual Disk Into Multiple Files selected to allow the disk to increase by increments of 2 GB and be easily defragmented.

8. Select Customize Hardware to change the RAM memory size if necessary.

9. The last step of the menu allows you to check the settings of the virtual machine (Figure 7.4). Click Finish to create the virtual machine.

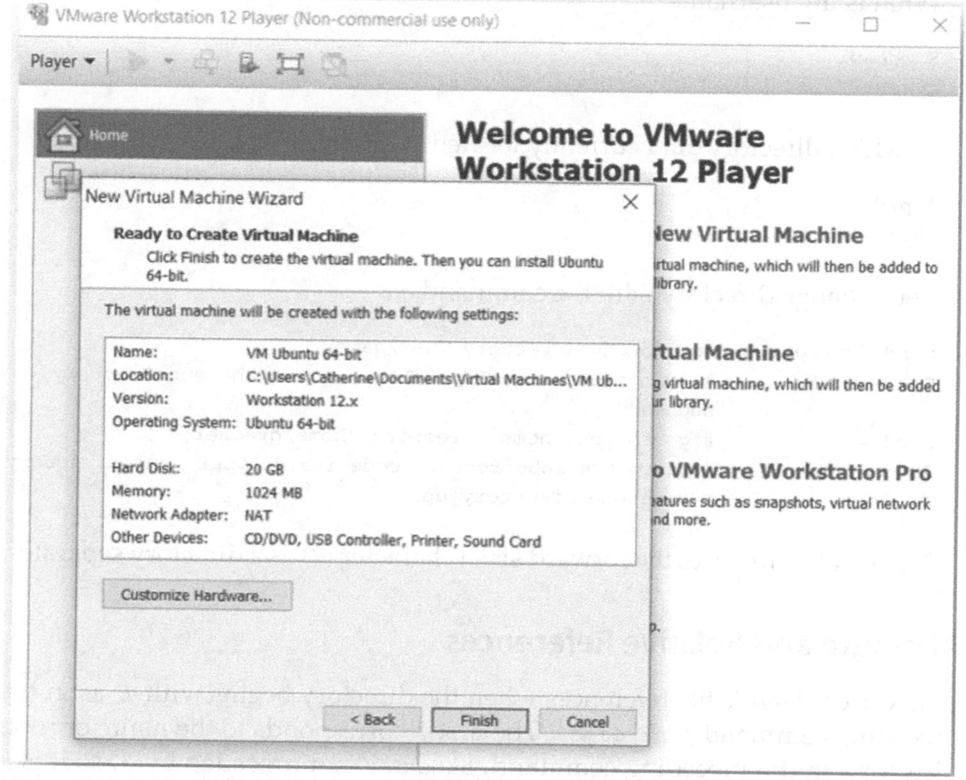

Figure 7.4: Configuration of an Ubuntu Virtual machine with VMware on Windows 10

Getting Familiar with the Command Line

Let's open a terminal to get access to the Linux command line. If you have a local Linux on your laptop (via native support, or dual-boot, or virtual machine) and want to practice the command line of your Linux laptop, you only need to right-click the Linux desktop background and select Open Terminal. To connect to a remote computer, such as an AWS instance, you need to first open a terminal (Linux terminal, Windows terminal via PuTTY, or MacOS terminal via its embedded terminal) and then log in remotely and securely via the ssh command.

The functionalities of the command line are provided by a program called shell. The most common shell in most Linux distributions is the Bourne Again shell (Bash). The dollar sign ($) is the prompt that indicates that the command line or shell is ready to accept commands. Let's now look at some useful commands to get to know our environment. Linux commands are often shortened by removing the vowels—for example, the commands "move" and "copy" are shortened to mv and cp, respectively.

What is my username?

```
$ whoami
cvacher
```

In which directory am I currently located?

```
$ pwd
/home/cvacher
```

Let's change directory with the command `cd`.

```
$ cd /usr/games   #go to the directory /usr/games
$ cd sudoku       # from the current directory, go to the subdirectory
                  sudoku
$ cd ~            # go to your home directory /home/cvacher
$ cd ~/code       # go to the subdirectory code inside your home directory
$ cd ..           # move one directory up
```

Note that Linux uses the forward slash character (/) as a directory separator.

Absolute and Relative References

Linux uses *absolute* file references when the directory begins with /, as in the preceding command /usr/games. The first / corresponds to the main, or root, directory. In the preceding command, sudoku is not preceded by /, so it is a *relative* file reference, meaning we request to go to the subdirectory sudoku from the current directory. The tilde character (~) refers to the home directory.

As a convenient shortcut, you can use the Tab key to automatically complete commands—a real time-saver. If you are in a directory that contains the following subdirectories:

```
abel
baker
charlie
```

. . . and you were to type this at the command line:

```
$ cd a<Tab>
```

. . . the command line would complete the command for you:

```
$ cd abel
```

That is because the command line is smart enough to predict what you want based on its knowledge of the subdirectories that exist. If you had subdirectories named `abel` and `axel`, you would have to remove the ambiguity by entering the first two characters of the name before hitting Tab.

```
$ cd ab<Tab>
```

Or:

```
$ cd ax<Tab>
```

The following are some common tasks you may want to perform.
To list the files contained in your current directory, use the `ls` command.

```
$ ls
mahjong sudoku
```

To list detailed information about the files contained in your current directory, use this:

```
$ ls -alh
-rwxr-xr-x  1 root root 158K Aug 25  2018 sudoku
-rwxr-xr-x  1 root root 105K Nov  2  2017 mahjong
```

Note the following options in this `ls` command:

- `-a` requests the most recent files to be listed first.

- `l` indicates detailed information, including file permissions, owner, group, size, and creation date.

- `-h` stands for human readable so that file sizes are given in units (KB, MB, GB) and therefore more readable.

To get help on a command, for example, the `ls` command, use the `man` (aka manual) function.

```
$ man ls
NAME     ls - list directory contents
SYNOPSIS ls [OPTION]... [FILE]...
DESCRIPTION ....
```

To visualize the content of a file, use the `cat` command.

```
$ cat myfile.txt
This is the content of myfile.txt.
```

If the file to visualize is very long, the following commands are very useful.
You can use the `more` command to scroll down a file one line at a time (press Enter) or one page at a time (press the spacebar).

```
$ more myfile.txt
```

The `less` command is similar, but more powerful, as it allows you to scroll up and down with the up/down arrows and PageUp/PageDown keys. Type q to exit. If the -s option is added, the lines that are longer than the terminal window are not wrapped, but truncated, which makes visualization easier.

```
$ less -S myfile.txt
```

You can use the `less` command to find search terms with -p (*p* for pattern). Type n to move to the next matching term, and type N to move to the previous matching term. There is no space between -p and the search item.

```
$ less -pSmith listofnames.txt
```

The `head` and `tail` commands print the first or last n lines of a file, respectively, where n is 10 by default.

```
$ head -30 myfile.txt    # visualize the first 30 lines of myfile.txt
$ tail -5 myfile.txt     # visualize the last 5 lines of myfile.txt
```

Manipulating Files

To create a new empty file, use the `touch` command.

```
$ touch test.txt
$ ls
   touch.test
```

To delete files, use the `rm` command.

```
$ rm test.txt
```

To copy a file, use the `cp` command.

```
$ cp file1.txt file2.txt
```

To rename a file, or move it to another directory, use the `mv` command.

```
$ mv file.txt newfile.txt     # rename file.txt into newfile.txt
$ mv newfile.txt newdir       # move newfile.txt to a new
                              directory newdir
```

To create and remove directories, use the `mkdir`, `rmdir`, and `rm -r` commands.

```
$ mkdir mynewdir
$ rmdir mynewdir
```

Note that the `rmdir` command will work only if the directory is empty! To remove recursively a directory as well as all its subdirectories and files, use the `rm -r` command.

```
$ rm -r mynewdir
```

Transferring Files to and from Your AWS Instance

You can easily transfer files between Linux machines or between Linux and macOS machines; you just need to use the `scp` (secure copy) command from a terminal. Supposing the IP address of your AWS instance is `170.160.10.15`, you can transfer files from your Linux or macOS laptop to the AWS instance with this:

```
$ scp myfile.txt 170.160.10.15:~/
```

and vice versa with the following:

```
$ scp 170.160.10.15:~/myfile.txt ./
```

To transfer files between a Windows laptop and the AWS instance, the easiest solution is to use a File Transfer Protocol (FTP) package such as WinSCP or FileZilla. Figure 7.5 shows the FileZilla interface, with the laptop files on the left and the server files on the right. You can copy and paste from one side to the other to securely transfer files.

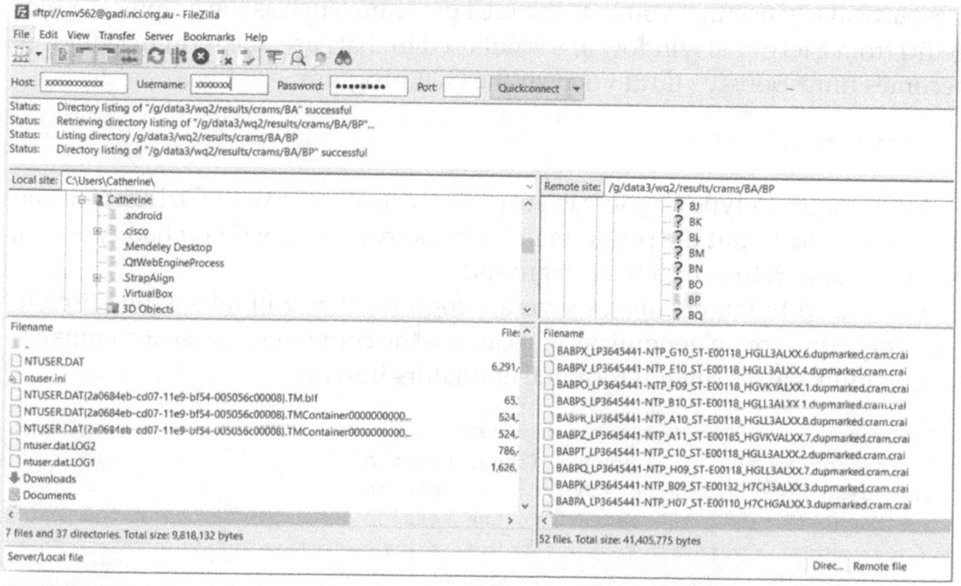

Figure 7.5: FTP client interface (FileZilla) that allows the transfer of files between laptop and AWS instance

In Windows text files, a new line is coded by the combination of two characters: a carriage return and a line feed. In Linux, as well as in the more recent macOS, a new line is coded only by a line feed. Therefore, a Windows text file transferred via FTP to a Linux server will need to be converted to the new format. This is easily done with the dos2unix utility.

```
$ dos2unix windows-textfile.txt converted-textfile.txt
```

The reverse command, unix2dos, can be used after transferring from Linux to Windows.

Keyboard Shortcuts

Using the command line involves typing long commands, as well as long filenames and directory names. Fortunately, some shortcuts are very handy.

- **Up and down arrows:** Move to your last used commands.
- **Tab key:** Autocompletes a line of text. For example, instead of typing **cd myfavouritedirectory**, only type **cd myf** and press Tab, and the myfavouritedirectory will automatically autocomplete.

Running Programs in the Background

Appending & to a command allows the command to be executed in the background. It is particularly useful if you launch a GUI program (such as the graphical editor gedit) from a terminal window. If & is not used in that case, the terminal window becomes unresponsive until you exit the GUI program.

```
$ gedit myfile.txt &
```

If you forget to type &, you can suspend the program with Ctrl+Z. You can then type in **bg** to put the program in the background (like if you had typed &), and type **fg** to return it to the foreground.

If you need to launch one or several programs that will take a long time to run (very common in genomics), you can use the convenient screen command. Screen multiplexes several terminal emulators into one.

```
$ screen    .       # start a new session
$ ./myprogram.sh.   # launch a (long) program
CTRL+a+d.           # detach from the session
$ screen -r.        # reattach to the session
```

If you have launched several sessions, you can list the available sessions and their process IDs.

```
$ screen -ls
```

You can then reattach to the session of your choice (process ID 12345, for example).

```
$ screen -r 12345
```

Understanding File Permissions

File permissions can be either read access, write access, or execution access. The permissions are granted at the level of the file owner (called user), group, and for all others.

Figure 7.6 indicates how the permissions are recorded.

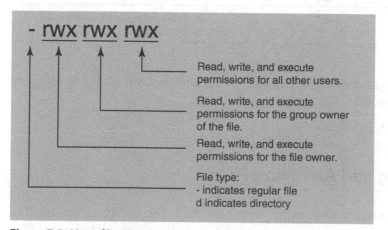

Figure 7.6: Linux file permissions

Let's look at the `ls` command we executed previously.

```
$ ls -alh
-rwxr-xr-x  1 root root 158K Aug 25  2018 sudoku
```

You can see that `sudoku` is a file (-) that has read/write/execute permissions for the root user (`rwx`), plus read and execute permissions for the group as well as for others (`r-x` repeated twice).

These permissions can be altered with the `chmod` command. The following command defines permissions to the file `myfile.txt`: read and write access to the file owner/user (u=rw), read access to the group (g=r), and no access to others (o=).

```
$ chmod u=rw,g=r,o= myfile.txt
```

These permissions can also be granted using the octal code, as shown in Table 7.2.

Table 7.2: Linux Permissions, Octal Code, and Interpretation

PERMISSIONS	OCTAL CODE	INTERPRETATION
rwxrwxrwx	777	Read, write, and execute for all users
rwxr-xr-x	755	Read and execute for all users; write permission for owner
rwx------	700	Read, write, and execute for owner; no permission for anybody else
rw-rw-rw-	666	Read and write permission for all
r--------	400	Read for owner; no permission for anybody else

The following command grants read, write, and execute access to all users for the myfile.txt file:

```
$ chmod 777 myfile.txt
```

Compressing and Archiving Files

The tar utility's name comes from "tape archiver." The utility can be used to archive a group of files to disk or other media. It is also often used to move files between disks or machines.

Let's archive and compress the proteinwork subdirectory of my home directory onto the disk /g/data.

```
$ tar cvfz /g/data/proteinwork.tgz ~/proteinwork
```

The -c option indicates that we are creating an archive and not extracting files from an archive. The -v option stands for verbose to show the tar progress. The -f option tells tar the name of the archive to create (proteinwork.tgz).

The -z option says we want to create a compressed archive with gzip. Note that the file extension of the compressed archive can be either .tgz or .tar.gz. To extract the archive, replace the option -c with the option -x (extract).

```
$ tar xvfz /g/data/proteinwork.tgz
```

tar can use bzip2 instead of gzip for compressing the archive. Bzip2 achieves about 15 percent higher compression rates than gzip but is slower. It can be invoked with the -j option. The file extension of the compressed archive can then be bz2, tbz, or tb2.

```
$ tar cvfj /g/data/proteinwork.bz2 ~/proteinwork   # archive and
                                                   compress with bzip2
$ tar -xvfj videos-14-09-12.tar.bz2.               # extract and
                                                   decompress
```

Compression

Compression is a computational technique that reduces the amount of space required to store a given amount of data. Usually, this is done by replacing long strings of bytes that appear many times in a data set with a small representative symbol that takes up much less storage space.

The Linux compression utilities gzip (decompression utility: gunzip; file extension .gz) and bzip2 (decompression utility: bunzip2, file extension .bz2) compress individual files.

```
$ bzip2 AAA.bam
```

The zip program is available on several operating systems and can create archives like tar.

```
$ zip myzip.zip file1.txt file2.txt   # compresses the 2 files into a
                                      single zip file myzip.zip
$ zip -r myzip.zip mydir.             # compresses the contents of
                                      directory mydir into
                                      # myzip.zip
```

Note that compressed files can be visualized without decompressing them, by using the zcat command. Other commands work on compressed files, such as zless, zmore, zgrep, or the utilities zdiff and zcomp, which compare files. These commands are referred to as Z *commands* and simplify our lives by decompressing files temporarily in the temporary storage directory /tmp.

```
$ zcat AAAA.fastq.gz
```

Grep

Very often in genomics work, you will want to pick lines and sections of interest out of large text files. For example, you might want to learn where a particular sequence of nucleotides appears in a whole-genome sequence.

`grep` is a useful utility that searches for text in a file and returns the line where there is a match.

```
$ grep HomoSapiens myfile.txt .        # Returns lines with the word
                                         HomoSapiens
$ grep -i HomoSapiens myfile.txt.      # Same, but makes the search
                                         case insensitive
```

We can do the opposite and search for lines without text.

```
$ grep -v HomoSapiens myfile.txt   # Returns lines without the word
                                     HomoSapiens
$ grep -c HomoSapiens myfile.txt   # Counts the number of lines with the
                                     word HomoSapiens
```

`Grep` can use Linux regular expressions, as follows:

- `^` represents the beginning of a line.
- `$` identifies the end of a line.
- Characters inside brackets, `[ ]`, constitute bracket expressions. For example, the bracket expression `c[aou]t` matches *cat*, *cot*, and *cut*.
- A range expression such as `c[5-7]t` matches *c5t*, *c6t*, and *c7t*.
- The dot (`.`) represents any single character except a new line.
- `?` corresponds to zero matches or one match.
- `+` represents one or more matches.

How do you easily count the number of sequences in a FASTA file? The FASTA sequences start with > at the beginning of a line, so `grep` counts the number of lines beginning with >.

```
$ grep -c "^>" reads.fa
```

Pipes and Redirection Operators

Pipes, represented by the character `|`, enable you to send the output of one program to another program as input. For example, the following command visualizes a BAM file; passes the output to the `cut` utility, which cuts the sixth column (to extract the CIGARs—that is, alignment reports for each read); and then sends the output to the `sort` utility and then to the `uniq` utility, which counts the number of unique CIGARs:

```
$ samtools view AAAA.bam | cut -f 6 | sort | uniq -c
```

Redirection operators can, for example, save the lengthy screen output (called *standard output*) of a program to a text file, which can be easily read later (see Table 7.3).

Table 7.3: Redirection Operators

REDIRECTION OPERATOR	ACTION
>	Creates a new file that contains the standard output
>>	Appends the standard output to the existing file
<	Sends the contents of the specified file to be used as standard input
<<	Accepts additional lines as standard input

In the following example, we visualize a compressed FASTQ file with `zcat`, extract the top 5,000 lines, recompress the result, and redirect the standard output to create a test FASTQ file. This can be used to test a pipeline on a small scale.

```
$ zcat AAAA.fq.gz | head -5000 | gzip > test_AAAA.fq.gz
```

The following command counts the number of reads in a FASTQ file:

```
$ zcat AAAA.fq.gz | echo $((`wc -l`/4))
```

The `wc` (word count) utility with the `-l` (line) option counts the number of lines. As each read in a FASTQ file is described by four lines, the result needs to be divided by 4.

Text Processing Utilities: awk and sed

For complex text processing tasks, Python and Perl are probably more appropriate than `awk` and `sed`. These two utilities can, however, be quite powerful in one or two lines of code.

`Awk` is actually more than a text processing utility. It is a scripting language that supports conditional statements, loops, variables, etc., like any other scripting language.

Awk can do the following:

- Read a file line by line
- Separate fields or columns on the basis of a delimiter
- Search for a pattern using regular expressions and take an action if there is a match

The syntax is typically as follows:

```
$ awk ' { some code } ' myfile.txt     # myfile is read line by line, and
                                        some code is executed
                                       # on each line.
```

$1 refers to the first field, $2 to the second, and so on, and FS represents the field separator.

To print to the screen the second column of myfile, using a tab as delimiter, use this:

```
$ awk '{ print $2}'  FS='\t' myfile.txt
```

There are two exceptions to the "execute the code line by line" rule: the BEGIN block is executed before myfile is read, and the END block is executed.

Let's print the average of the first column of a file.

```
$ awk 'BEGIN {x=0}  {x=x+$1;}  END {print x/NR}'  myfile.txt
```

The BEGIN block is used to initialize x to 0, the file is then read line by line, and x is incremented by the value of the first column at each line. Finally, in the END block, x is divided by NR, where NR is a special awk variable representing the row number.

sed can be best defined as an automated text editor. The most common sed command is the substitution command, where in the following example the term *cold* in the old file is replaced by the term *hot* in the new file. The redirection operators < and > are defined in the next section.

```
$ sed s/cold/hot/ < oldfile.txt  >newfile.txt
```

Sed is very convenient for deleting lines.

```
$ sed '1,3d'
```

Managing Linux

In Linux, *root* refers to the administrator or superuser. The root account is useful for performing administrative tasks such as installing new software, managing user accounts, or maintaining disks. There are three ways to acquire root privileges.

- Log in as root.
- Type the su command to change your identity to root ($ su).
- Type the sudo command in front of the command you want to execute ($ sudo command).

Note that in all three cases, you will be prompted for the root password.

Package Management Systems

As previously mentioned, the Linux distributions contain a package management system, which maintains a database of all the installed packages. The two main package managers are Debian (Debian and Ubuntu distributions) and RPM.

The set of commands of the Debian package manager are as follows:

```
$ sudo apt upgrade                  # upgrade all installed packages to
                                    their latest versions
$ sudo apt install package_name.    # install a new package
$ sudo apt remove package_name.     # remove a package
$ sudo apt purge package_name.      # remove package and
                                    configuration files
```

RPM uses a different set of commands.

```
$ sudo yum upgrade                  # upgrade all installed packages to
                                    their latest versions
$ sudo yum install package_name     # install a new package
$ sudo yum remove package_name      # remove a package, including its
                                    configuration files
```

The AWS Command-Line Interface

As you gain familiarity with the browser-based AWS Console, you may come to regard it as more of a hindrance than an aid to efficient work and productivity. You may find yourself wishing for a more straightforward way to do what you want, without a lot of mouse movements and clicking. You will want a command line, and you will be delighted to learn that AWS has an excellent one.

The AWS Command-Line Interface (CLI) is a set of tools that runs either on your local computer—Windows, macOS, or Linux—or on a remote computer dedicated to the role of AWS administration. You install the AWS CLI software package and are then able to use hundreds of text commands that enable you to do pretty much anything you want to do with AWS.

On Windows, you run AWS CLI commands at the traditional command line (cmd.exe) or in PowerShell. In a Unix-like environment, such as Linux or macOS, you run commands from within your favorite shell—Bash, zsh, tcsh, or another one. There is also a way to run a form of AWS CLI commands from within the AWS Console, using an AWS service called Systems Manager.

After gaining some familiarity with the AWS CLI, you will find that it is more time-efficient to perform most tasks there. You will also come to appreciate that you can use the AWS CLI in command-line scripts. That makes it easy to replicate the performance of AWS jobs quickly and repeatably.

Installing the AWS CLI Environment

Before you can use the AWS CLI, you have to install it on your workstation (your laptop computer or the EC2 instance that will serve as your administrative workstation). Setting up the CLI is a straightforward matter, described here.

Windows

To install the CLI under Microsoft Windows, find the installer (.msi) that suits your version of the operating system, either 32 or 64 bit. Run the installer as you would any other Windows setup wizard.

In 64-bit Windows environments, the CLI installs to C:\Program Files\ Amazon\AWSCLI.

In 32-bit Windows environments, the CLI installs to C:\Program Files (x86)\ Amazon\AWSCLI.

To confirm that the installation completed successfully, go to the Windows command line (cmd.exe) or PowerShell and enter the following:

```
aws --version
```

The command line should return information about the versions of AWS, Python, and Windows itself, something like this:

```
aws-cli/2.0.12 Python/3.7.4 Windows/7
```

macOS and Linux

To install the AWS CLI under Linux, Unix, or macOS, go to the operating system's command line and enter this:

```
curl "https://s3.amazonaws.com/aws-cli/awscli-bundle.zip" -o "awscli-bundle.zip"
```

This downloads the AWS CLI configuration package from AWS. Then, use the unzip command to unzip the downloaded file.

```
unzip awscli-bundle.zip
```

With that done, you can enter this last command to actually run the installation script:

```
sudo ./awscli-bundle/install -i /usr/local/aws -b /usr/local/bin/aws
```

Note that the sudo command will prompt you for the machine's administrative password.

Having done that, and with the installation of the AWS CLI apparently complete, you can verify that it works as expected with the aws command.

```
aws --version
```

The command line should return information about the versions of AWS, Python, and Windows itself, something like this:

```
aws-cli/2.0.12 Python/3.7.4 Darwin/19.5.0 botocore/2.0.0dev16
```

Configuring the AWS CLI

The most basic task is connecting to the AWS instance you want to operate upon.

You will recall that the Access Key ID and Secret Access Key were created when you set up your account and configured IAM roles and users. You set up users and also set up programmatic access, the latter of which is where Access Key IDs and Secret Access Keys came into the picture. You should have saved both values in a safe location when you created them. If you did not, you will need to create a new Access Key ID and matching Secret Access Key. That is because the Secret Access Key cannot be viewed, other than one time immediately after it is created. If you do not have the Secret Access Key for the account, you want to access via the AWS CLI. Go back into Identity and Access Management (IAM). There, you should delete any key sets for which you have lost the Secret Access Keys (since those combinations cannot be used) and create a new one for your use at the AWS CLI.

Setting the Configuration at the Command Line

You can make changes to the setup of the AWS command line with the `aws configure` command, which will prompt you for the AWS Access Key ID and AWS Secret Access Key matching your target instance. The basic command sequence is as follows:

```
aws configure
AWS Access Key ID [None]: <Your Access Key ID>
AWS Secret Access Key [None]: <Your Secret Access Key>
Default region name [None]: <The name of the AWS region you want to work
in, such as ap-southeast 2>
Default output format [None]: ENTER
```

This puts you at the AWS command line, logged in and ready to operate on your AWS resources. If you find yourself operating on more than one instance frequently (say, for your professional work as well as your personal activities), it is possible to set up a configuration file that lets you choose from among several sets of access credentials.

Storing the Configuration in the Configuration File

If you prefer not to set your security configuration at the command line, you can instead manually edit the file in which your credentials are stored.

Windows

In Windows, your security configuration information is stored in the User Profile directory. The path to the directory is contained in the `%USERPROFILE%` environment variable, which the system administrator can set to pretty much anything. By default, though, that directory is as follows:

```
C:\Users\default
```

or as follows:

```
C:\Users\<Username>
```

In the `%USERPROFILE%` directory, there will be a folder called `aws`, within which is a file called `credentials`. The file that holds the access keys are in this file.

Linux and macOS

In Linux and macOS, open the following file in your favorite text editor:

```
~/.aws/config
```

Find the section that looks like this:

```
[default]
aws_access_key_id = <access key id>
aws_secret_access_key = <secret access key>
region = ap-southeast-2
```

Edit the file to include your Access Key ID and Secret Access Key. Once you have done that, use chmod to set correct access to the file:

```
chmod 600 ~/.aws/config
```

This gives read and write access to you, and you only.

Testing Your Installation

An easy way to verify that your AWS CLI installation has worked is to run a simple command that neither creates nor destroys anything. By going to your workstation's command line and entering the following, you will get a listing of all the S3 buckets in the instance you've logged into:

```
aws s3 ls
```

If that does not work, go back and investigate your CLI setup. The problem most likely has to do with permissions.

AWS CLI Essentials

When using the AWS CLI, the general idea is that you go to your workstation's command line (which is to say, the shell in macOS or Linux, or the Windows command line or Power Shell) and invoke the aws command, followed by sub-commands and arguments that specify what you want to do.

One of the design goals the architects of the AWS CLI seem to have had in mind is consistency across client platforms. In other words, an AWS command sequence that works when connected via a Windows computer should also work when you are using a Linux machine to connect to the AWS cloud. This allows for portability of your skills across client platforms and makes it possible to share shell scripts that use the AWS command set among users of different kinds of client machines.

There are some minor differences among the AWS CLI implementations for different operating systems. For example, in Windows, it is necessary to enclose certain values in double quotes (" "), while the Linux implementation expects single quotes (' ').

AWS CLI commands are nested. In other words, the leading commands determine which following commands you can use. All command sequences begin with the following:

```
aws
```

After that, you specify which AWS service you want to work with. For example, you would enter the following to work with DynamoDB databases:

```
aws dynamodb
```

And you would enter the following to work on EC2 instances:

```
aws ec2
```

From there, you specify further subcommands, followed by parameters that designate specifically how the string of commands and subcommands are to be carried out. To start an EC2 instance, for example, you would enter the following:

```
aws ec2 start-instances --instance-ids <The instance ID(s) of the
instance(s) you want to start>
```

Once you are into the AWS CLI environment, you can use the `help` command anywhere you are not sure what to enter next. Entering `help` rather than an "actual" command or parameter should give you a listing of all possible options, with a short explanation of what each option does.

An Alternative Approach: AWS Systems Manager

If you prefer not to install the AWS CLI on your local machine or if you do not have one available for some reason, you can use AWS Systems Manager instead. There is a lot more to this tool—you will find that documentation refers to it as AWS SSM for some reason—than the command line, but it does allow you to log into the AWS Console and issue many of the same commands you would use at a local instance of the AWS CLI.

To access AWS SSM, go to the AWS Console and choose it from the list of available services. From the SSM main page, choose the Quick Setup option, which will identify all EC2 instances, and certain other AWS resources, that are associated with your AWS account.

On the left side of the SSM environment, choose Run Command and then click the orange Run Command button. Choose the command you want to run from the list of available commands—it is not as comprehensive as the set of commands available from the AWS CLI package you can install on your own machine—and specify any required parameters. Click Run, and observe as SSM monitors the command as it runs.

Summary

In this chapter, you got a taste of how the Linux command line works. That's handy because you will be doing a lot of your bioinformatics work there. In particular, we began to explore the AWS command-line interface (CLI). You will use that to interact directly with AWS services at the command line and include AWS-specific commands in repeatable and adaptable scripts.

The next chapter explores specific bioinformatics workflows, and we begin to perform genomics work on the AWS cloud.

Processing the Sequencing Data

In this chapter, we run a pipeline to convert sequencing data into gene information. The sequencing data consists of millions of short reads generated by Illumina or BGI short read DNA sequencing. The gene information includes the DNA variants and mutations found in the genes of the person whose DNA was sequenced. When the variant is only 1 nucleotide long, it is called a *single nucleotide variation* (SNV) or *single nucleotide polymorphism* (SNP). Variants that are 2 to 50 nucleotides long are called *indels*. Anything bigger than that is called a *structural variant* (SV) or *copy number variant* (CNV).

Getting from Data to Information

To get from sequencing data to gene information, we can run the following steps, which make up our data processing pipeline:

1. Align the sequencing reads to the reference genome to produce a BAM file.
2. Make adjustments and refinements to the aligned reads in the BAM file.
3. Identify the small differences (SNVs and indels) in this data compared to the reference genome and record them in the VCF file.
4. Make adjustments and refinements to the variants in the VCF file.

5. Annotate the SNVs and indels so that we will know whether they are inside a gene and whether the consequence of the variant might be deleterious.

6. Prioritize the variants to be able to identify the most consequential ones.

7. When analyzing a family or trio (mother, father, and child), carry out inheritance analysis to see which variants are in family members that have a disease and are not in the other members of the family who don't have the disease.

8. Identify and annotate SVs and CNVs.

Let's look at software commands to process these eight steps, step-by-step. And at each step of the pipeline, let's see what a particular variant of interest would look like at that stage of the pipeline. We will follow an example variant in the cystic fibrosis gene CFTR. The variant is p.Phe508del, which causes the phenylalanine amino acid (Phe), normally at position 508 of the amino acid sequence of the protein, to be deleted from the CFTR protein. It is a deletion of 3 nucleotides (CTT) at position `chr7:117559591-117559593` in the GRCh38 reference genome, which is `chr7:117199645-117199647` in the GRCh37 reference genome. In the sequence of coding DNA nucleotides, this deletion is c.1521_1523del.

The wild-type sequence consists of the following nucleotides, which translate into a chain of amino acids starting with Proline (Pro) at position 499:

- **Nucleotides:** CCTGGCACCATTAAAGAAAATATCAT**CTT**TGGTGTTTCCTATGAT

- **Amino acids:** ProGlyThrIleLysGluAsnIle**IlePhe**GlyValSerTyrAsp

- **Position of amino acids:**
  499500501502503504505506507**508**509510511512513

The variant-containing sequence resulting from the deletion of Phe508 is as follows:

- **Nucleotides:** CCTGGCACCATTAAAGAAAATATCATTGGTGTTTCCTATGAT

- **Amino acids:** ProGlyThrIleLysGluAsnIleIleGlyValSerTyrAsp

- **Position of amino acids:**
  499500501502503504505506507509510511512513

The CTT sequence in the wild-type sequence spans two codons. You may notice that the Phe amino acid at position 508 is no longer present. You may also note that the codon for the Ile at position 507 has changed from ATC to ATT, which both code for isoleucine, and so the Ile507 remains.

Aligning to the Reference Genome

As a result of sequencing, we have hundreds of millions of sequencing reads in the FASTQ files, each one from a different place in the genome. We need to align those reads to the reference genome. It is like putting the pieces of a jigsaw puzzle into their correct places, where the guiding outlines of the correct places are the reference genome.

For each sequenced read, there are four lines in the FASTQ file. The first part of the first line is the sequence identifier. The actual DNA sequence is recorded on the second line. The third line is a simple separator with the plus sign. The fourth line has the same length as the second line because it is the quality information about nucleotide in the sequence. Here is what one of the paired reads might look like:

```
@A01234:A0:HM1A2ABC1:4:1101:1054:1000 1:N:0:AAGAGGCA+NCGCATAA
GTGGAAGAATTTCATTCTGTTCTCAGTTTTCCTGGATTATGCCTGGCACCATTAAAGAAAATATCATTGGTG
TTTCCTATGATGAATATAGATACAGAAGCGTCATCAAAGCATGCCAACTAGAAGAGGTAAGAAACTATGTGA
AAACTT
+
FFFFFFFF:FFFFFFFFFFFFFFFFFFFFFFFFFFFFFFFFFFFFFFFFFFFFFFFFFFFFFFFFFFFFFFFF
FFFFFFFFFFFFFFFFFFFFFFFFFFFFFFFFFFFFFFFFFFFFFFFFFFFFFFFFFFFFFFFFFFFFFFFFFF
FF:,FF
```

As mentioned in Chapter 4, paired reads sequencing involves sequencing DNA fragments from each end of the fragment. Paired reads have the same sequence identifier and are stored in separate FASTQ files traditionally named with the suffixes R1.fastq and R2.fastq.

As you can imagine, the FASTQ files are large and therefore are stored in a compressed format so that they take up less disk space. Most programs reading FASTQ files can read the compressed files. If your FASTQ file is not already compressed and you want to compress it, you can use the bgzip command from the htslib library. The code for the htslib library is at github.com/samtools/htslib. To obtain this software, click the Releases link, and then right-click the link for the latest release and copy the link address. Download that latest software release.

```
wget https://github.com/samtools/htslib/releases/download/1.14/
htslib-1.14.tar.bz2
```

Decompress the htslib software and go into the newly created directory of this software.

```
tar -xf htslib-1.14.tar.bz2
cd htslib-1.14
```

Then follow the instructions in the INSTALL file to install the htslib software.

```
sudo apt-get update # Ensure the package list is up to date
sudo apt-get install autoconf automake make gcc perl zlib1g-dev libbz2-
dev liblzma-dev libcurl4-gnutls-dev libssl-dev
./configure
make
make install
```

If your FASTQ files are plain text and you'd like to compress them so that they take up less room on disk, you can now do so with the bgzip command from the htslib library.

```
bgzip my_fastq.R1.fastq
bgzip my_fastq.R2.fastq
```

This will replace those files with their compressed versions: my_fastq .R1.fastq.gz and my_fastq.R2.fastq.gz. You can use the less command to view their uncompressed content and Ctrl+C when you are finished looking with the less command.

```
less -S my_fastq.R1.fastq.gz
```

The FASTQ files are usually delivered in compressed format, so you won't need to run the bgzip command to compress them. If, for some reason, you want to uncompress the fastq.gz file, then use the gunzip command:

```
gunzip my_fastq.R1.fastq.gz
```

We will align the FASTQ reads to the reference genome to produce a BAM file. The program we will use is BWA, so let's download and install the BWA software.

```
git clone https://github.com/lh3/bwa.git
cd bwa
make
```

After creating the alignment file, we will compress it with the Samtools software, so let's get that software too. Samtools is a suite of programs to interact with high-throughput sequencing data. Go to github.com/samtools/samtools. Click the Releases link, right-click the link for the latest release, and copy the link address. Download the latest software release.

```
wget https://github.com/samtools/samtools/releases/download/1.16.1/
samtools-1.16.1.tar.bz2
```

Decompress the Samtools software and go into the newly created directory of this software.

```
tar -xf samtools-1.16.1.tar.bz2
cd samtools-1.16.1
```

Then follow the instructions in the INSTALL file to install the Samtools software.

```
./configure
make
make install
```

To align these reads to a reference genome, we need to get the reference genome. Specifically, let's get the human reference genome version GRCh38 (also known as hg38), because it is more complete compared to its predecessor GRCh37 (also known as hg19, and successor of hg18). There are two main types of GRCh38. One version has multiple alternative regions defined, called *alt-contigs*, for some of the genomic regions where there are a lot of differences found in different people. The other version, the *no-alt-contigs* version, doesn't have these multiple redefinitions. To avoid the issues that arise with the alt-contigs reference, we will use the no-alt-contigs reference. Let's download it.

```
wget
ftp://ftp.ncbi.nlm.nih.gov/genomes/all/GCA/000/001/405/GCA_000001405.15_
GRCh38/seqs_for_alignment_pipelines.ucsc_ids/GCA_000001405.15_GRCh38_no_
alt_analysis_set.fna.gz
```

The programs that will be using this reference genome require various indexes, so let's create them.

```
gunzip GCA_000001405.15_GRCh38_no_alt_analysis_set.fna.gz
bwa index GCA_000001405.15_GRCh38_no_alt_analysis_set.fna
samtools faidx GCA_000001405.15_GRCh38_no_alt_analysis_set.fna
```

So now we have our pair of FASTQ files, a reference genome, the BWA alignment, and the Samtools software. Here are the BWA and Samtools commands to align FASTQ reads to the reference genome, with the output immediately piped through to Samtools to compress the output:

```
outfile=my_bam.bam
infile1=my_fastq.R1.fastq.qz
infile2=my_fastq.R2.fastq.gz
ref_fasta=GCA_000001405.15_GRCh38_no_alt_analysis_set.fna
NCPUS=4 # To reduce execution time, use as many CPUs as you have available
ID= HM1A2ABC1:4 # Illumina platform unit, may contain flowcell and lane
PU=HM1A2ABC1:4:MY_SAMPLE_1 # Illumina platform unit, may contain
flowcell, lane, and sample
LB=20200101 # library prep id
PL=ILLUMINA # platform technology used for sequencing
```

```
SM=MY_SAMPLE_1 # sample identifier
read_group_id="@RG\tID:$ID\tPU:$PU\tLB:$LB\tPL:$PL\tSM:$SM" # This will
be attached to every read in the output so that downstream quality
improvement algorithms will know where the read came from

bwa mem -t $NCPUS -R "$read_group_id" "$ref_fasta" "$infile1" "$infile2" | \
    samtools view -1 - -o $outfile
```

The BWA output is immediately piped through to Samtools to compress it, so we don't spend time and disk space creating an intermediate text file of the alignments; instead, we can go straight to producing a compressed version. Given that it is compressed, we need a decompression tool such as Samtools to see the plain-text output, which consists of header records followed by one line for each read alignment. Press Ctrl+C when you have finished looking at the output from the less command.

```
samtools view -h my_bam.bam | less -S
```

If you want to spend time and disk space creating an intermediate text file of alignments called mysam.sam, then run the following commands instead of the previous commands:

```
outfile1=my_sam.sam
outfile2=my_bam.bam
infile1=my_fastq.R1.fastq.gz
infile2=my_fastq.R2.fastq.gz
ref_fasta=GCA_000001405.15_GRCh38_no_alt_analysis_set.fna
NCPUS=4
read_group_id="@RG\tID:$ID\tPU:$PU\tLB:$LB\tPL:$PL\tSM:$SM"

bwa mem -t $NCPUS -R "$read_group_id" "$ref_fasta" "$infile1" "$infile2"
> $outfile1

samtools view -1 $outfile1 -o $outfile2
```

The intermediate output file is in plain-text format, so you can inspect it with any of these commands:

```
head my_sam.sam
tail my_sam.sam
less -S my_sam.sam
more my_sam.sam
```

It is recommended to only work with BAM files and use samtool with the view option to visualize BAM files.

Perhaps you have already received a BAM file from the sequencing provider rather than the FASTQ files. If you would like to use this BAM file for further processing, some of that processing may require the presence of the same reference genome file that was used to create the BAM file and that was probably not provided with the BAM file. If you would like to realign your sequencing reads to the same reference genome that we are using in these exercises, you can extract your reads from the BAM file as FASTQ. You can then align the FASTQ reads to a reference genome, as we have just done. To extract the reads from a BAM as FASTQ, we need the Bedtools software in addition to Samtools, so let's download and install Bedtools. Go to `github.com/arq5x/bedtools2/` `releases`, right-click the bedtools.static.binary link, select Copy Link Address, and run the following:

```
wget https://github.com/arq5x/bedtools2/releases/download/v2.30.0/
bedtools.static.binary
ln -s bedtools.static.binary bedtools
```

Now let's extract FASTQ read pair files from an existing BAM file. We use Samtools to sort the BAM file by sequence read identifier so that read pairs will be one after the other. We use Samtools to filter the BAM file reads so that we get each read pair only once and don't get any extra copies that are "not primary alignment" or "supplementary alignment." Then we use Bedtools to extract the read pairs, with the first read of a read pair going into the first FASTQ output file and the second read of a read pair going into the second FASTQ output file. The output FASTQ files are plain text, so we then compress them to take up less disk space.

```
infile=my_bam.bam
outfile1=my_fastq.R1.fastq
outfile2=my_fastq.R2.fastq

samtools sort -n "${infile}" | \
    samtools view -F 2304 | \
    bedtools bamtofastq -i - -fq "${outfile1}" -fq2 "${outfile2}"

bgzip $outfile1
bgzip $outfile2
```

If you received more than one pair of FASTQ files, you align each pair separately to its own BAM file and then merge the BAM files during the MarkDuplicates step of the next section.

After aligning the FASTQ reads to the GRCh38 reference genome and generating a BAM file, the tab-delimited entry for our example sequence will be the following:

```
A01234:A0:HM1A2ABC1:4:1101:1054:1000   99   chr7      117559533  60.  67M3D83M
        =    117559783    400
GTGGAAGAATTTCATTCTGTTCTCAGTTTTCCTGGATTATGCCTGGCACCATTAAAGAAAATATCATTGGTGTT
TCCTATGATGAATATAGATACAGAAGCGTCATCAAAGCATGCCAACTAGAAGAGGTAAGAAACTATGTGAAAACTT
FFFFFFFF:FFFFFFFFFFFFFFFFFFFFFFFFFFFFFFFFFFFFFFFFFFFFFFFFFFFFFFFFFFFFFFFFFFF
FFFFFFFFFFFFFFFFFFFFFFFFFFFFFFFFFFFFFFFFFFFFFFFFFFFFFFFFFFFFFFFFFFFFFFFFFF:,FF
```

As mentioned in Chapter 4, the tab-delimited entry indicates the sequence identifier (identical for the two paired reads, so this read and its mate), the chromosome and position in the chromosome this read maps to, the same information for this read's mate, a CIGAR field, a flag, a mapping quality score, and the length of the original fragment.

Here, our example sequence maps to chromosome 7 at position `117559533`, and its mate maps to the same chromosome (the = field) at position `117559783`. The CIGAR 67M3D83M indicates that 67+83=150 nucleotides perfectly match (M) the genome reference with a 3bp deletion (D) in the middle. The flag 99 means this read is the first read of the pair and both reads are mapped. The mapping quality score of 60 (top score) indicates excellent mapping quality, and the fragment size is estimated to be 400 nucleotides, which is in the normal range.

The definition of BAM file fields is found on page 6 of the BAM file specification at `samtools.github.io/hts-specs/SAMv1.pdf`. The flag field values are explained at `broadinstitute.github.io/picard/explain-flags.html`. Enter the `sam_flag` value, and let that website explain that particular value.

Making Adjustments and Refinements to the Aligned Reads in the BAM File

We broadly follow the best practices developed and recommended by the Broad Institute and use the popular Genome Analysis ToolKit (GATK) software to process the aligned sequencing reads.

Let's download and install the GATK software. Go to `github.com/broadinstitute/gatk/releases`, right-click the gatk zip file link to copy the link address, and then download the GATK software package.

```
wget https://github.com/broadinstitute/gatk/releases/download/4.3.0.0/
gatk-4.3.0.0.zip
unzip gatk-4.3.0.0.zip
cd gatk-4.3.0.0
less -S README.md
```

You will see in the `README.md` file that GATK requires Java version 8. If you don't already have Java 8, then please install it.

```
sudo apt install openjdk-8-jre
```

The best practices recommend performing the MarkDuplicates step. It is possible for the one sequenced fragment to be sequenced more than once, as there are PCR duplicates or optical duplicates. The MarkDuplicates step identifies these extra reads and marks them as duplicates so that downstream processing can take that into account. Why would we want this taken into account? One example is when the software is calling a variant and has to decide whether half the reads contain the variant and half the reads contain the reference genome value, as would be expected if the variant is on one chromosome but not on the other. If there are extra PCR or optical duplicates for one of the reads containing the variants, then if the software doesn't know that they are duplicates, it may erroneously conclude that the variant is on both chromosomes when really it is on only one of the chromosomes.

The following commands run the MarkDuplicates step. If you have more than one input BAM due to having had more than one pair of FASTQ input files, then this is the place to merge them by entering them all as input to the MarkDuplicates step.

```
infile=my_bam.bam
outfile=my_bam_markdup.bam

java -server -jar gatk-package-4.3.0.0-local.jar \
    MarkDuplicates \
    INPUT="$infile" \
    OUTPUT="$outfile" \
    METRICS_FILE=output_MarkDuplicatesMetrics.txt \
    VALIDATION_STRINGENCY=SILENT \
    OPTICAL_DUPLICATE_PIXEL_DISTANCE=2500 \
    ASSUME_SORT_ORDER="queryname" \
    MAX_FILE_HANDLES=1000 \
    MAX_RECORDS_IN_RAM=5000000 \
    CLEAR_DT=false \
    ADD_PG_TAG_TO_READS=false

# If there is more than one input bam file, then replace the above INPUT
line with:
# INPUT=my_bam1.bam INPUT=my_bam2.bam INPUT=my_bam3.bam
```

Check that the MarkDuplicates step has worked correctly.

```
cat output_MarkDuplicatesMetrics.txt
```

If the MarkDuplicates step has worked correctly, the output of the preceding cat command will be as follows:

```
No errors found
```

If the MarkDuplicates step has worked correctly, you can delete its input to free up disk space.

```
rm my_bam.bam
Now sort the output bam of MarkDuplicates by chromosome and co-
ordinates, so
that it can be processed by the SetNmMdAndUqTags step:
infile=my_bam_markdup.bam
outfile=my_bam_markdup_sort.bam

gatk SortSam \
    --INPUT "$infile" \
```

```
    --OUTPUT "$outfile" \
    --SORT_ORDER "coordinate" \
    --COMPRESSION_LEVEL 2 \
    --CREATE_INDEX false \
    --CREATE_MD5_FILE false
```

Next, in the SetNmMdAndUqTags step, we set additional tags in the BAM file, which will be used in downstream processing.

```
infile=my_bam_markdup_sort.bam
outfile=my_bam_markdup_sort_tags.bam
ref_fasta=GCA_000001405.15_GRCh38_no_alt_analysis_set.fna

gatk SetNmMdAndUqTags \
    --INPUT "$infile" \
    --OUTPUT "$outfile" \
    --CREATE_INDEX true \
    --CREATE_MD5_FILE true \
    --REFERENCE_SEQUENCE "$ref_fasta"
```

Let's validate our latest version of the BAM file to make sure it is still in good shape.

```
infile=my_bam_markdup_sort_tags.bam

java -server -jar gatk-package-4.3.0.0-local.jar \
    ValidateSamFile \
    I="$infile" \
    MAX_OPEN_TEMP_FILES=1000 \
    MODE=VERBOSE > output_Picard_ValidateSamFile.txt

cat output_Picard_ValidateSamFile.txt
```

If the SortSam and SetNmMdAndUqTags steps have worked correctly, the output of the preceding cat command will be as follows:

```
No errors found
```

If the SortSam and SetNmMdAndUqTags steps have worked correctly, their input files can be deleted in order to free up disk space.

```
rm my_bam_markdup.bam my_bam_markdup_sort.bam
```

The next major refinement to make to our BAM alignment file of sequencing reads mapped to a reference genome is the GATK Base Quality Score Recalibration (BQSR). This process applies machine learning to improve the nucleotide-specific quality scores. This is useful, because downstream processes rely on those quality scores to help decide whether to call variants. For instance, if there are a few reads showing a variant in a given position but their quality scores are low, the downstream processing may decide to *not* call the variant, given that the quality scores say that we should not be confident about whether the nucleotide really is the recorded value. There are two steps to BQSR: building a model by comparing the existing BAM file with known true positive variants, and applying the model to produce a new BAM file having adjusted nucleotide quality scores.

The BQSR process requires some reference files for building the BQSR models. To find those files, search for "GATK resource bundle," which will bring you to `gatk.broadinstitute.org/hc/en-us/articles/360035890811-Resource-bundle`, which will direct you to `console.cloud.google.com/storage/browser/genomics-public-data/resources/broad/hg38/v0`. You will be required to log in with a Google account to see the files on this web page. The files we need for BQSR are dbsnp, known_indels, and Mills_and_1000G_indels—more specifically:

```
Homo_sapiens_assembly38.dbsnp138.vcf

Homo_sapiens_assembly38.dbsnp138.vcf.idx

Homo_sapiens_assembly38.known_indels.vcf.gz

Homo_sapiens_assembly38.known_indels.vcf.gz.idx

Mills_and_1000G_gold_standard.indels.hg38.vcf.gz

Mills_and_1000G_gold_standard.indels.hg38.vcf.gz.idx
```

You can download those files from your browser and transfer them to your Linux environment where we are running the pipeline. If you want to download directly to Linux, then you will need to obtain the gsutil software to do that. Go to `cloud.google.com/storage/docs/gsutil_install` and follow the instructions for installing gsutil.

```
sudo apt-get install apt-transport-https ca-certificates gnupg
echo "deb [signed-by=/usr/share/keyrings/cloud.google.gpg]
https://packages.cloud.google.com/apt cloud-sdk main" | sudo tee -a /
etc/apt/sources.list.d/google-cloud-sdk.list
curl https://packages.cloud.google.com/apt/doc/apt-key.gpg | sudo apt-
key --keyring /usr/share/keyrings/cloud.google.gpg add -
sudo apt-get update && sudo apt-get install google-cloud-sdk
gcloud init
```

When the gsutil software is working, you can download the reference files for BQSR with the following commands:

```
/hg38/v0/Mills_and_1000G_gold_standard.indels.hg38.vcf.gz.tbi . gsutil
cp gs://genomics-public-data/resources/broad/hg38/v0/Homo_sapiens_
assembly38.dbsnp138.vcf .
gsutil cp gs://genomics-public-data/resources/broad/hg38/v0/Homo_
sapiens_assembly38.dbsnp138.vcf.tbi .
gsutil cp gs://genomics-public-data/resources/broad/hg38/v0/
Homo_sapiens_assembly38.known_indels.vcf.gz .
gsutil cp gs://genomics-public-data/resources/broad/hg38/v0/
Homo_sapiens_assembly38.known_indels.vcf.gz.tbi .
```

```
gsutil cp gs://genomics-public-data/resources/broad/hg38/v0/Mills_
and_1000G_gold_standard.indels.hg38.vcf.gz .
gsutil cp gs://genomics-public-data/resources/broad
```

Now that you have the reference files required for BQSR, let's run the two steps of BQSR.

```
infile=my_bam_markdup_sort_tags.bam
outfile=my_bam_bqsr.bam
ref_fasta=GCA_000001405.15_GRCh38_no_alt_analysis_set.fna
out_bqsr_table=my_bam_bqsr.table

dbsnp=Homo_sapiens_assembly38.dbsnp138.vcf.gz
known_indels=Homo_sapiens_assembly38.known_indels.vcf.gz
gold_std_indels=Mills_and_1000G_gold_standard.indels.hg38.vcf.gz

# BQSR Step 1 - building a model by comparing the existing bam file with
known true positive variants
gatk BaseRecalibrator \
    -R "$ref_fasta" \
    -I "$infile" \
    --use-original-qualities \
    --known-sites "$dbsnp" \
    --known-sites "$known_indels" \
    --known-sites "$gold_std_indels" \
    -O "$out_bqsr_table"

# BQSR Step 2 - applying the model to produce a new BAM file having
adjusted nucleotide quality scores
gatk ApplyBQSR \
    -R "$ref_fasta" \
    -I "$infile" \
    --bqsr-recal-file "$out_bqsr_table" \
    --static-quantized-quals 10 \
    --static-quantized-quals 20 \
    --static-quantized-quals 30 \
    --add-output-sam-program-record \
    --create-output-bam-index \
    --create-output-bam-md5 \
    --use-original-qualities \
    -O "$outfile"
```

Again, let's validate this latest BAM file that we have created, and if no errors are found, we can delete the input BAM file to BQSR to free up disk space.

```
infile=my_bam_bqsr.bam

java -server -jar gatk-package-4.3.0.0-local.jar \
    ValidateSamFile \
    I="$infile" \
    MAX_OPEN_TEMP_FILES=1000 \
    MODE=VERBOSE > output_Picard_ValidateSamFile.txt
```

```
cat output_Picard_ValidateSamFile.txt
# If no errors found, then the output will be: No errors found
rm my_bam_markdup_sort_tags.bam
```

The BQSR process has created an index for the sorted BAM file, so we can use Samtools to look at a specific region of the BAM file to see the reads mapped there. Now that BQSR has been run, you may see that the nucleotide-level quality scores have changed for that read in the CFTR gene. Here is the samtools command to look at that region:

```
samtools view my_bam_bqsr.bam chr7:117199645-117199647
```

The quality scores (PHRED scores) in the 11th tab-delimited field have changed compared to what was originally in the FASTQ file for this read, as in the following example of our CFTR read:

```
A01234:A0:HM1A2ABC1:4:1101:1054:1000    99      chr7      117559533   60
67M3D83M =       117559783   400

GTGGAAGAATTTCATTCTGTTCTCAGTTTTCCTGGATTATGCCTGGCACCATTAAAGAAAATATCATTGGTG
TTTCCTATGATGAATATAGATACAGAAGCGTCATCAAAGCATGCCAACTAGAAGAGGTAAGAAACTATGTG
AAAACTT

5??????+??????????????????5?????5+??????????????????5?5??'????????+??+555?
5???'??????????5?+?5????????5?5??5??++5+???'????????'5???5'????5?????'
+5?5+??+??
```

Identifying the Small Differences and Recording Them in the VCF File

Now that all sequencing reads are aligned to the human reference, we need to call the variants (SNVs, indels) at each nucleotide position. If the sequencing depth is 30x, around 30 reads cover each nucleotide position. If about 50 percent (respectively 100 percent) of reads differ from the human reference at a nucleotide position, we can call a heterozygous (respectively homozygous) SNV. If about 50 percent of reads have a CIGAR that reports a 3 nucleotides deletion (text "3D" in the CIGAR), as in the CFTR deletion example, a heterozygous small deletion indel should be called.

In the following commands, GATK software programs will be run to study the BAM file and call SNVs and indels. The SNV and indel variants' results will be reported in the output VCF file. GATK programs include extra steps to assess if variants are likely to be sequencing errors instead of genuine variants. Calling the variants of several samples in a single step, known as *joint-calling*, can improve accuracy. In our example, we have only one sample running through the multistep best practices pipeline. However, this pipeline also allows joint-calling multiple samples.

Using the BAM file as input, let's create an intermediate file called the gVCF file.

```
infile=my_bam_bqsr.bam
outfile=my_gvcf.gvcf
ref_fasta=GCA_000001405.15_GRCh38_no_alt_analysis_set.fna
gatk_dbsnp=Homo_sapiens_assembly38.dbsnp138.vcf.gz

gatk HaplotypeCaller \
    -R "$ref_fasta" \
    --dbsnp "$gatk_dbsnp" \
    -I "$infile" \
    -ERC GVCF \
    -O "$outfile"
```

Using the gVCF file as input, let's create the intermediate GenomicsDB database.

```
infile=my_gvcf.gvcf
output_directory=my_genomics_db_directory
ref_fasta=GCA_000001405.15_GRCh38_no_alt_analysis_set.fna

gatk GenomicsDBImport \
    -V "$infile" \
    --genomicsdb-workspace-path "$output_directory" \
    --reader-threads 4

# If this is a joint-call and there is more than one input sample,
# then replace the above -V line with:
# -V my_gvcf1.gvcf -V my_gvcf2.gvcf -V my_gvcf3.gvcf
```

Using the intermediate GenomicsDB database as input, let's call the SNV and indel variants.

```
input_directory=my_genomics_db_directory
outfile=my_vcf.vcf
ref_fasta=GCA_000001405.15_GRCh38_no_alt_analysis_set.fna
gatk_dbsnp=Homo_sapiens_assembly38.dbsnp138.vcf.gz

gatk GenotypeGVCFs \
    -R ${ref_fasta} \
    -O ${outfile} \
    -D ${gatk_dbsnp} \
    -G StandardAnnotation \
    --only-output-calls-starting-in-intervals \
    --use-new-qual-calculator \
    -V gendb://${input_directory}
```

The output from this part of the pipeline is a VCF file containing SNVs and variants. The definition of VCF file fields is found in samtools.github.io/hts-specs/VCFv4.2.pdf. For each variant, VCF records its chromosome (CHROM) and position (POS), its identifier in the dbsnp database if available (ID), the value of the nucleotide(s) in the reference genome (REF) and in the observed BAM file (ALT), and its genotype ("0/1" if heterozyggous and "1/1" if homozygous).

You can inspect the output with the less command (and you end the command by pressing Ctrl+C). After the header rows that start with a # symbol, you will see the variant rows, with one row per variant.

```
less -S my_vcf.vcf
```

Here is what our CFTR variant would look like in the VCF file:

```
#CHROM POS        ID           REF    ALT QUAL   FILTER INFO FORMAT   MY_SAMPLE
chr7   117559591 rs113993960 TCTT   T   1000   PASS   GT              0/1
```

Making Adjustments and Refinements to the Variants in the VCF File

Now that all of the SNVs and indel variants have been called, further GATK software analyses can help determine if variants are false-positive sequencing errors or true-positive variants. This is the Variant Quality Score Recalibration (VQSR) process. The seventh column of variant rows in the VCF file is the FILTER field, and it contains the "PASS" value for variant calls that are considered to be true positives. In the VQSR process, the variants are ranked in order of confidence that they are true positives, and the variants at the bottom of the pile have their "PASS" value changed to something else so that we know that the analysis found that these variants may not be true positives after all. This process is carried out for SNVs and indels separately. It uses some reference files from the "GATK resource bundle" that we previously downloaded and used in the BQSR process.

The first step of VQSR involves using machine learning to build a model for our variant data, using known high-quality reference data.

```
infile=my_vcf.vcf
ref_fasta=GCA_000001405.15_GRCh38_no_alt_analysis_set.fna
mills_resource_vcf= Mills_and_1000G_gold_standard.indels.hg38.vcf.gz
axiomPoly_resource_vcf= Axiom_Exome_Plus.genotypes.all_populations.poly.
hg38.vcf.gz
dbsnp_resource_vcf= Homo_sapiens_assembly38.dbsnp138.vcf.gz
hapmap_resource_vcf= hapmap_3.3.hg38.vcf.gz
omni_resource_vcf= 1000G_omni2.5.hg38.vcf.gz
one_thousand_genomes_resource_vcf= /1000G_phase1.snps.high_confidence.
hg38.vcf.gz

gatk VariantRecalibrator \
    -V ${infile} \
    -O my_output_indel.recal \
```

```
    --tranches-file my_outfile_indel.tranches \
    --rscript-file my_outfile_indel_rscript_plots.R \
    -an FS -an ReadPosRankSum -an MQRankSum -an QD -an SOR -an DP \
    -mode INDEL \
    --resource:mills,known=false,training=true,truth=true,prior=12
${mills_resource_vcf} \
    --resource:axiomPoly,known=false,training=true,truth=false,prior=10
"${axiomPoly_resource_vcf} \
    --resource:dbsnp,known=true,training=false,truth=false,prior=2
${dbsnp_resource_vcf}

gatk VariantRecalibrator \
    -R ${ref_fasta} \
    -V ${infile} \
    -O my_output_SNP.recal \
    --tranches-file my_outfile_SNP.tranches \
    --rscript-file my_outfile_SNP_rscript_plots.R \
    -an QD -an MQ -an MQRankSum -an ReadPosRankSum -an FS -an SOR \
    -mode SNP \
    --resource:hapmap,known=false,training=true,truth=true,prior=15
${hapmap_resource_vcf} \
    --resource:omni,known=false,training=true,truth=true,prior=12
${omni_resource_vcf} \
    --resource:1000G,known=false,training=true,truth=false,prior=10
${one_thousand_genomes_resource_vcf} \
    --resource:dbsnp,known=true,training=false,truth=false,prior=7
${dbsnp_resource_vcf}
```

The second step of VQSR involves applying the machine learning model to identify variants in our data that may not be true positives after all and changing their filter field so that they no longer have the value of "PASS."

```
infile=my_vcf.vcf
outfile=my_vcf_vqsr.vcf

gatk ApplyVQSR \
    -V ${infile} \
    -O my_intermediate_output.vcf \
    --recal-file my_output_indel.recal \
    --tranches-file my_outfile_indel.tranches \
    --truth-sensitivity-filter-level 99.7 \
    -mode INDEL

gatk_path ApplyVQSR \
    -V my_intermediate_output.vcf \
    -O ${outfile} \
    --recal-file my_output_SNP.recal \
    --tranches-file my_outfile_SNP.tranches \
    --truth-sensitivity-filter-level 99.7 \
    -mode SNP
```

You now have a VCF file containing SNV and indel variants that has gone through best-practices processing so that we can be as confident as possible that the reported variants really are genuine variants (and not erroneous artifacts) while not missing any variants. You can inspect this output with the `less` command. After the header rows that start with a # symbol, the variants are listed with one row per variant.

```
less -S my_vcf_vqsr.vcf
```

You may notice that sometimes the variant's ALT field, which is the fifth field, has more than one value, with each value separated by a comma. This means that there is more than one variant at that particular position of the chromosome (the chromosome and position are the first and second fields, respectively). In the next section, we will annotate each variant with general information about the variant, such as its consequence and rarity or frequency in the general population. That information will be inserted into the info field of the VCF record, which is the eighth field. Different variants will have different consequences and population frequencies. Thus, we need to decompose those multi-allelic variants to be multiple mono-allelic variants so that each mono-allelic variant is on its own row. Annotation involves comparing our variants to known variants. However, sometimes there can be multiple ways of representing a variant. For example, the sequence at `chr7:117603759` is CAAC. Let's say that our sample has a 1 base-pair deletion: CAC. Is it the first A or the second that is deleted? Is this variant represented as `chr7:117603759CA>C` (left-aligned), or is it represented as `chr7:117603760AA>A` (right-aligned)? The convention is to report variants as left-aligned. Thus, before comparing our variants to others for annotation purposes, let's ensure that our variants are left-aligned.

The software tool we will use for decomposing and left alignment normalizing our variants is vt. Let's install vt.

```
git clone https://github.com/atks/vt.git
cd vt
git submodule update --init --recursive
make
make test
```

Here is the code to decompose the multi-allelic variants:

```
infile=my_vcf_vqsr.vcf
outfile=my_vcf_vqsr_dcmps.vcf.gz

bgzip $infile
tabix -p vcf "${infile}".gz
vt decompose -o $outfile -s "${infile}".gz
tabix -p vcf $outfile
```

And here is the code to carry out the left-alignment normalization:

```
infile=my_vcf_vqsr_dcmps.vcf.gz
outfile=my_vcf_vqsr_dcmps_nrml.vcf.gz

vt normalize -o $outfile -s $infile
tabix -p vcf "$decompose_vcf"
```

Annotating the SNVs and Indels

We now have our sample's variants—SNVs and indels—recorded in the VCF file. As shown in Chapter 4, each human has around six million SNVs and indels when compared to the reference genome. Most of them are not problematic. We need to annotate the variants to identify which variants occur near or inside a gene or in important regulatory regions. Annotations also document the consequences of these variants, such as change of amino acid (*missense* variant), truncation of the protein (*stopgain* variant), or complete change in protein sequence (*frameshift* variant, where an indel changes the reading frame of codons). Other important annotations include frequency of the variant in the general population (usually derived from the EXAC database of 60,000 individuals), pathogenicity score (predicted by machine learning methods), and information from the clinical ClinVar database ("Likely pathogenic," "Likely benign," or "Variant of unknown significance").

VEP is a well-documented, frequently updated software for annotating human variants in a VCF file. The base VEP installation provides some annotations, whereas others are provided by VEP plugins. Installation instructions for VEP are available at m.ensembl.org/info/docs/tools/vep/script/vep_download .html, and we follow them here to install VEP.

First, some Perl modules are required to run VEP, so let's install them.

```
cpan DBI
cpan archive::Zip
cpan DBD::mysql
cpan Bio::DB::HTS
```

Now install VEP.

```
git clone https://github.com/Ensembl/ensembl-vep.git
cd ensembl-vep
perl -v # Make sure that you have perl version 5
perl INSTALL.pl --CACHEDIR /my/vep/cache
```

The VEP installation process sometimes pauses to ask you a question about the species data you want to install. In those cases, choose homo-sapiens and GRCh38. At one point, the choice of homo-sapiens GRCh38 is between VEP, *refseq*, and *merged*. Choose *merged* so as to have the most comprehensive annotations. The VEP installation process also asks you if you want to install plugins. Reply yes to those requests.

Although we have replied yes to installing the code for VEP plugins, some of the plugins require installing additional reference files. Here are instructions for obtaining VEP plugin data for plugins that we will use:

```
cd vep_plugin_data
#
# Get CADD reference files
wget https://krishna.gs.washington.edu/download/CADD/v1.6/GRCh38/whole_
genome_SNVs.tsv.gz
wget https://krishna.gs.washington.edu/download/CADD/v1.6/GRCh38/whole_
genome_SNVs.tsv.gz.tbi
#
# Get REVEL reference files
wget https://www.google.com/url?q=https%3A%2F%2Frothsj06.u.hpc.mssm.
edu%2Frevel-v1.3_all_chromosomes.zip&sa=D&sntz=1&usg=AFQjCNHuQrC
vuq--CoiCGwtDvQhBnTYj2Q
unzip revel-v1.3_all_chromosomes.zip
cat revel_all_chromosomes.csv | tr "," "\t" > tabbed_revel.tsv
sed '1s/.*/#&/' tabbed_revel.tsv > new_tabbed_revel.tsv
bgzip new_tabbed_revel.tsv
tabix -f -s 1 -b 2 -e 2 new_tabbed_revel.tsv.gz
#
# Get ExAC reference files
wget http://ftp.ensembl.org/pub/data_files/homo_sapiens/GRCh38/
variation_genotype/ExAC.0.3.GRCh38.vcf.gz
wget http://ftp.ensembl.org/pub/data_files/homo_sapiens/GRCh38/
variation_genotype/ExAC.0.3.GRCh38.vcf.gz.tbi
```

Now that VEP and some interesting VEP plugins are installed, let's run VEP on our VCF file so as to obtain annotations for our variants.

```
infile=my_vcf_vqsr_dcmps_nrml.vcf.gz
outfile=my_vcf_vep.txt

perl vep/ensembl-vep/vep --tab --format vcf --cache --merged --everything \
  --offline -i $infile -o $outfile \
  --dir_cache /my/vep/vep_cache --dir_plugins vep/ensembl-vep \
  --use_given_ref --force_overwrite --assembly GRCh38 \
  --plugin CADD,vep_plugin_data/whole_genome_SNVs.tsv.gz \
  --plugin REVEL,vep_plugin_data/new_tabbed_revel.tsv.gz \
  --custom ExAC.0.3.GRCh38.vcf.gz,gnomADg,vcf,exact,0,ExAC_AF
```

The output can be loaded into Excel to view as a spreadsheet. There are often multiple different transcript possibilities, each with its own set of consequences. VEP outputs a line for each of them, and thus one input variant can result in multiple output lines of annotation. Some of the output columns are of particular interest. In the VEP consequence column, a value of start_lost is rather serious if it occurs in an important gene transcript, because it means that the transcript will not be translated into a protein. A `splicing` variant is also likely to be harmful if this is an important gene. A value of `intron_variant` is probably of no consequence to the protein. In the ExAC column, a value equal to zero or below 1 percent population frequency corresponds to a rare variant. Pathogenic variants are usually rare, although the opposite is not necessarily true. In the CADD and REVEL columns, high predicted pathogenicity scores indicate a higher likelihood that the variant is harmful.

Prioritizing the Variants to Identify the Most Consequential Ones

As each individual has around six million variants, a big challenge is to prioritize the variants. Many prioritization methods are available. We will use the publicly available and easy-to-use VPOT software.

The input to VPOT is traditionally a VCF file, so let's run the VEP annotation again so as to produce an annotated VCF file.

```
infile=my_vcf_vqsr_dcmps_nrml.vcf.gz
outfile=my_vcf_vep.vcf

perl vep/ensembl-vep/vep --vcf --format vcf --cache --merged --everything \
   --offline -i $infile -o $outfile \
   --dir_cache /my/vep/vep_cache --dir_plugins vep/ensembl-vep \
   --use_given_ref --force_overwrite --assembly GRCh38 \
   --plugin CADD,vep_plugin_data/whole_genome_SNVs.tsv.gz \
   --plugin REVEL,vep_plugin_data/new_tabbed_revel.tsv.gz \
   --custom ExAC.0.3.GRCh38.vcf.gz,gnomADg,vcf,exact,0,ExAC_AF
```

The annotated VCF output of VEP is in a nonstandard format, whereas VPOT and all other VCF-reading software requires a VCF input in standard VCF format. Let's convert the nonstandard VEP VCF to standard VCF format.

```
infile=my_vcf_vep.vcf
outfile=my_vep_vcf.vcf

python3 VPOT/VPOT_6_utility.py -i $infile -o $outfile
```

Let's install the VPOT software.

```
git clone https://github.com/VCCRI/VPOT.git
```

The VEP software allows you to specify the scores to give to various values in the info field subfields of your VCF file. The info field is the eighth field, and after the annotation step, it contains values for each annotation subfield, for each variant. The list of annotation subfields in the info field is listed in the header of the VCF (in the lines that commence with ##INFO=<ID=). For each variant, VPOT will assign the appropriate score for each info subfield and then add up all the scores to give a final score for the variant. The output of VPOT is the variants sorted and ranked by their VPOT variant score. Here is a recommended VPOT Prioritization Parameter File (PPF) for our VCF file that was annotated by VEP earlier. The values in this file are tab-delimited.

```
cat my_ppf.txt
PF   ExAC_ALL   0.01   LE
PD   CADD_PHRED   N   -998   1   10   0   20.000001   2   20   4
VT   VEP_Consequence   splicing   60
VT   VEP_Consequence   splice_acceptor_variant   60
VT   VEP_Consequence   splice_donor_variant   60
VT   VEP_Consequence   stop_gained   100
VT   VEP_Consequence   transcript_ablation   100
VT   VEP_Consequence   start_lost   100
VT   VEP_Consequence   frameshift_variant   75
VT   VEP_Consequence   frameshift_deletion   75
VT   VEP_Consequence   frameshift_insertion 75
VT   VEP_Consequence   stop_lost   40
VT   VEP_Consequence   non_coding_transcript_variant   -28
VT   VEP_Consequence   intron_variant -28
GN   Gene Symbol
GN   SYMBOL
```

The preceding PF line causes the VPOT output to be filtered to include only variants whose ExAC frequency value is less than or equal to 0.01 (1 percent of the population). If you want all variants to be prioritized and output by VPOT, then remove that line in the PPF file.

VPOT also requires another input file listing the VCF file and the sample identifier in the VCF file. The sample identifiers in a VCF file are found in the header row.

```
#CHROM POS ID REF ALT QUAL FILTER INFO FORMAT MY_SAMPLE_1   MY_SAMPLE_2
MY_SAMPLE_3
```

Here is the content of this additional input file for our scenario:

```
cat my_vpot_list.txt
my_vep_vcf.vcf.     MY_SAMPLE_1
```

Now we actually run VPOT.

```
infile=my_vep_vcf.vcf
out_prefix=my_vpot_priority_output
infile_list=my_vpot_list.txt
param_file=my_ppf.txt

python3 VPOT/VPOT.py priority $out_prefix $infile_list $param_file
```

The VPOT output filename will be the prefix we specify plus a random number to ensure that the output is unique and doesn't overwrite any previous VPOT output for this input VCF. The VPOT output file contains the filtered, prioritized variants, sorted in order of consequence according to the VPOT scoring that we specified in the PPF file, with the most harmful variants appearing first in the file.

Trio Analysis and Inheritance Analysis

Sequencing trios, such as parents and child, provide greater analysis power as we can select the variants present in the individual(s) with a medical problem and exclude the variants from unaffected family members. This is a powerful way to narrow down on candidate variants causal for a medical problem.

Bioinformatics teams have different methods for carrying out trio analysis. The VPOT software is an easy-to-use and easy-to-run software that can carry out inheritance modeling trio analysis on a VCF file containing variants of family members.

The VPOT software can analyze the following inheritance models:

- **Autosomal dominant (AD):** The child and one of the parents are affected by the genetic disease, while the other parent is not affected. This inheritance model occurs when the gene variant is able to cause the disease even when it is present on only one of the chromosomes and not on both chromosomes. It can be autosomal dominant either on the mother's side (the mother has the variant and the disease) or on the father's side (the father has the variant and the disease). Huntington's disease is an infamous example of an autosomal dominant inheritance pattern.

- **Autosomal recessive (AR):** The child is affected, but neither of the parents are affected; however, both parents are carriers. This inheritance model occurs when the gene variant is able to cause the disease only when it is present on both chromosomes. The parents each have one copy of the variant, and their other chromosome does not. The child, when inheriting one copy of the gene from the mother and the other copy from the father, unfortunately inherited the variant copy from each parent. Usually this variant renders the gene nonfunctional, so the child has no functioning copy of the gene, thus causing the disease. The parents each have one functioning copy and one defective copy of the gene, so they don't have the disease. Cystic fibrosis is an infamous example of the autosomal recessive inheritance pattern.

- **Compound heterozygous (CH):** This is similar to autosomal recessive except that it involves two different variants in the same gene. The parents each have one variant in a given gene that renders the gene nonfunctional, but they don't have the same nonfunctional variant. The parents are unaffected because they each have one remaining good copy of the gene. The child, however, has unfortunately inherited the variants from the parents, resulting in two different defective variants in the same gene, resulting in the child being affected by the disease. Cystic fibrosis is also an example of the compound heterozygous inheritance pattern.

- **De-novo (DN):** In this inheritance pattern, a de-novo variant has spontaneously arisen in the embryo resulting in the child having a deleterious variant that neither parent has. The child is affected by the disease while the parents are not. There are 40 to 80 de-novo variants in each embryo and thus in each newborn child. Usually they occur in a region of the genome that doesn't cause a problem, and thus most newborns are not born with a genetic disease. Sometimes, however, a child is unlucky in that a de-novo variant does cause a disease.

- **Case-control (CC):** In this scenario, there are family members with the disease and family members without the disease, and thus a genetic mutation is running through the family. Those family members unlucky enough to receive one copy of the variant are affected by the disease. In this analysis, VPOT lists all the variants that are in all of the affected individuals and are not in any of the unaffected individuals. This analysis is carried out when we have the sequenced variants of other family members, who may or may not be affected. We may or may not have the parents' sequenced variants to include in this analysis.

If you are analyzing a trio case where the child is affected and the parents are not, then VPOT inheritance models for AR, CH, and DN are appropriate. If you are analyzing a trio where one of the parents is affected in addition to the child, then the AR model is appropriate. If you have an assortment of affected and unaffected members from an extended family (perhaps the parents' DNA is not available, but DNA from cousins is available), then the VPOT CC analysis is helpful.

To carry out VPOT inheritance modeling, we need to specify the pedigree, that is, how the sequenced individuals are related to each other, and who is affected by the disease. VPOT's pedigree file format is the same as that used for the GATK software. It is tab-delimited, and the columns are the following:

- Family ID
- Individual ID
- Paternal ID (if you don't have the paternal sample for this individual, then put ND)
- Maternal ID (if you don't have the maternal sample for this individual, then put ND)
- Sex (1=male; 2=female; other=unknown)
- Phenotype (1=unaffected; 2=affected)

If you have sequencing data for the mother and father in addition to the child, then all three samples will have the same family ID. Make sure that the mother's individual ID is in the maternal ID field of the child's row and that the father's individual ID is in the paternal ID field of the child's row.

The input to the VPOT inheritance modeling process is a VPOT priority output file, created by running VPOT priority with an input file consisting of a joint-called VCF containing all the samples in the family, in the one VCF file. The individual IDs in the pedigree file must be set to the sample IDs that were in the VCF #CHROM record. When running the autosomal dominant (AD) inheritance model, the phenotype of one of the parents in the pedigree file must be set to 2 for affected.

```
infile=my_vpot_priority_output12345678.txt
out_prefix=my_vpot_inheritance_output_
pedigree_file=my_pedigree.ped
proband=MY_SAMPLE_1
inheritance_model=DN # possible values are DN, AD, AR, or CH

python3 POT/VPOT.py samplef $out_prefix $infile $pedigree_file $proband
$inheritance_model
ls -l "${out_prefix}${proband}"_"${inheritance_model}"_variant_filtered_
output_file_*.txt # this will display the file name of the output file
```

The filename of the output file is a composite of the `out_prefix` that you specified with the proband, the inheritance model, `variant_filtered_output_file`, and a random number suffixed to it.

The following runs VPOT inheritance with the case-control model. The filename of its output file will be the `out_prefix` that you specify with `variant_filtered_output_file` and a random number suffixed to it.

```
infile=my_vpot_priority_output12345678.txt
out_prefix=my_vpot_inheritance_output_
pedigree_file=my_pedigree.ped

python3 VPOT/VPOT.py samplef $out_prefix $infile $pedigree_file
ls -l "${out_prefix}CaseControl_variant_filtered_output_file_*.txt #
this will display the file name of the output file
```

The VPOT inheritance output contains the annotations as colon-delimited text in the info column. Let's split those annotations into separate tab-delimited columns so that the VPOT inheritance output file can be opened as a spreadsheet in a spreadsheet program.

```
infile=my_vpot_inheritance_output_MY_SAMPLE_1_DN_variant_filtered_
output_file_56781234.txt # or my_vpot_inheritance_output_CaseControl_
variant_filtered_output_file_56781234.txt
outfile=my_vpot_inheritance_tab_delimited_output.txt

sed -e 's/;/\t/g' $infile > $outfile
```

Now open the output file `my_vpot_inheritance_tab_delimited_output` `.txt` in a spreadsheet, either Microsoft Excel or LibreOffice Calc, to look at the results. All the variants and only the variants that satisfy the specific inheritance model will be in these results, sorted by VPOT score in descending order so that the more deleterious variants are at the top of the file. Ideally you will see the pathogenic variant causing the child's disease staring back at you as one of the first rows in this output file, with high pathogenicity scores, with a low or zero population frequency, and in a gene already known to be associated with the disease!

Identifying and Annotating SVs and CNVs

Structural variants (SVs) are larger variants than SNVs and indels and are usually defined as variants of more than 50 base pairs in length. They can be hundreds or thousands of nucleotides in length. As shown in Chapter 4, SVs can be called by two methods.

- Identifying breakends, where one DNA sequence is connected to a DNA sequence located in a nonadjacent region. This method can identify SVs at the nucleotide level.

- Spotting variations in read depth to detect copy number variants (CNVs) such as copy number loss (deletions) or copy number gain (duplications).

Many programs implement these techniques. Manta is a popular software for identifying breakends to call SVs. CNVnator (and the related CNVpytor from the same laboratory) are frequently used to detect CNVs, while Conanvarvar can spot very large CNVs of more than 1 million nucleotides in length.

Let's install and run Manta to identify structural variants in our sample. Here is how to install Manta:

```
apt-get update -qq
apt-get install -qq bzip2 gcc g++ make python zlib1g-dev
git clone https://github.com/Illumina/manta.git
cd manta
mkdir build && cd build
# Ensure that CC and CXX are updated to target compiler if needed, e.g.:
#       export CC=/path/to/cc
#       export CXX=/path/to/c++
../manta-${MANTA_VERSION}.release_src/configure --jobs=4 --prefix=/path/
to/install
make -j4 install
```

The following is how to run Manta. If you'd like to joint-call the SVs for three samples, replace the –bam line with --bam my_bam1.bam --bam my_bam2.bam --bam my_bam3.bam.

```
# Run Manta to call SVs

infile=my_bam_bqsr.bam
ref_fasta=GCA_000001405.15_GRCh38_no_alt_analysis_set.fna
out_directory=my_manta_output
outfile_vcf_gz="${out_directory}"/results/variants/diploidSV.vcf.gz
outfile_vcf="${out_directory}"/results/variants/diploidSV.vcf
outfile_with_inversions_vcf="${out_directory}"/my_manta_output.vcf

manta/build/bin/configManta.py \
   --bam $infile \
   --referenceFasta $ref_fasta \
   --runDir $out_directory

"${out_directory}"/runWorkflow.py --mode=local

# Run Manta to call Inversions

gunzip -f "${outfile_vcf_gz}"
manta/build/bin/convertInversion.py $link_to_samtools_bin $ref_fasta
$outfile_vcf > $outfile_with_inversions_vcf
```

We can run VEP on the Manta structural variant output to find out what regions of what genes are impacted by these structural variants. There is no point in including the SNV and indel plugin annotations because they will never match any of the structural variants. The output will be a tab-delimited file that can be viewed in a spreadsheet program.

```
infile=my_manta_output.vcf
outfile=my_manta_vep.txt

perl vep/ensembl-vep/vep --tab --format vcf --cache --merged
--everything \
   --offline -i $infile -o $outfile \
   --dir_cache /my/vep/vep_cache --dir_plugins vep/ensembl-vep \
   --use_given_ref --force_overwrite --assembly GRCh38
```

Let's install and run CNVnator to identify copy number variants in our sample. Here is how to install CNVnator and its prerequisite "root" software and specific Samtools package:

```
git clone https://github.com/root-project/root.git
mkdir root-build
cd root-build
cmake ..
make -j8
source bin/thisroot.sh
```

```
# CNVnator needs samtools with htslib incorporated in it
# Since CNVnator has been written, samtools was split into separate
packages for just samtools and just htslib
# However, CNVnator requires the samtools package that contains htslib
mkdir -p samtools_with_htslib/samtools-1.10
git clone https://github.com/samtools/htslib.git
git clone https://github.com/samtools/samtools.git
cd samtools_with_htslib/samtools-1.10
mkdir samtools_build
./configure --prefix=samtools_with_htslib/samtools-1.10/samtools_build
make all all-htslib
make install install-htslib

cd CNVnator_install
git clone https://github.com/abyzovlab/CNVnator.git
cd CNVnator

export ROOTSYS=root-build
ln -s samtools_with_htslib/samtools-1.10 samtools
make LIBS="-lcrypto"

export PATH=CNVnator_install/CNVnator/src:$PATH # do this when using
CNVnator or put this in .profile
```

Make sure the installation of the "root" software has worked, because CNVnator relies on this working.

```
source root/bin/thisroot.sh
root
.q
```

If you run into problems with CNVnator installation because of the root prerequisite, note that you can use instead CNVpytor, which is an extension of CNVnator built on Python. CNVpytor was created by the same laboratory as CNVnator and uses a similar engine and similar commands. The installation and usage of CNVpytor is described in Chapter 12, "Cancer Genomics." Now run CNVnator to identify CNVs in our sample. CNVnator does not have a joint-call feature. Each sample is run separately.

```
infile=my_bam_bqsr.bam
outfile_prefix=my_cnvnator_output
outfile=my_cnvnator_results.txt
ref_fasta=GCA_000001405.15_GRCh38_no_alt_analysis_set.fna

rm -rf "${outfile_prefix}".cnvnator_output.root
# Extract read mapping
CNVnator_install/CNVnator/cnvnator -root "${outfile_prefix}".cnvnator_
output.root -tree "${infile}" -chrom chr1 chr2 chr3 chr4 chr5 chr6 chr7
chr8 chr9 chr10 \
    chr11 chr12 chr13 chr14 chr15 chr16 chr17 chr18 chr19 chr20 chr21
chr22 chrX chrY chrM
```

```
# Generate histogram
CNVnator_install/CNVnator/cnvnator -root "${outfile_prefix}".cnvnator_
output.root -his 1000 -fasta "${ref_fasta}"

# Calculate statistics
CNVnator_install/CNVnator/cnvnator -root "${outfile_prefix}".cnvnator_
output.root -stat 1000

# Partition
CNVnator_install/CNVnator/cnvnator -root "${outfile_prefix}".cnvnator_
output.root -partition 1000

# Call CNVs
CNVnator_install/CNVnator/cnvnator -root "${outfile_prefix}".cnvnator_
output.root -call 1000 > $outfile

# https://github.com/abyzovlab/CNVnator/blob/master/README.md
# The output columns are as follows:
# CNV_type coordinates CNV_size normalized_RD e-val1 e-val2 e-val3 e-val4 q0
# where,
# normalized_RD -- read depth normalized to 1.
# e-val1 -- is calculated using t-test statistics.
# e-val2 -- is from the probability of RD values within the region to be in
the tails of a gaussian distribution describing frequencies of RD values
in bins.
# e-val3 -- same as e-val1 but for the middle of CNV
# e-val4 -- same as e-val2 but for the middle of CNV
# q0 -- fraction of reads mapped with q0 quality
```

The CNVnator output contains obvious CNVs where there is a large change in depth, as well as CNVs where there is little change in depth. We are not interested in the latter, so filter them out.

```
infile=my_cnvnator_results.txt
outfile=my_filtered_cnvnator_results.txt

echo -e "#Chr\tstart\tend\tcnv_type\tcnv_size\tnormalized_read_depth\
teval1_ttest\teval2_gauss_prob\teval3_ttest_midCNV\teval4_gauss_prob_
midCNV\tfraction_reads_mapped_with_q0_quality" > header.txt
cat $infile | awk 'BEGIN {FS="\t";OFS="\t"} {if (($5<0.05) && ($6<0.05)
&& ($7<0.05) && ($8<0.05)) {print $2,$1,$3,$4,$5,$6,$7,$8,$9}}' | sed
's/-/:/' | awk 'BEGIN {FS=":";OFS="\t"} {print $1,$2,$3}' | cat header.
txt - | cut -d$'\t' -f1-11 > $outfile
```

The CNVnator output is a tab-delimited file. To annotate it with VEP, convert it to a VCF file.

```
infile= my_filtered_cnvnator_results.txt
outfile=my_cnvnator_results.vcf
:>$outfile
```

```
echo '##fileformat=VCFv4.2' >> $outfile
echo '##FILTER=<ID=PASS,Description="All filters passed">' >> $outfile
echo '##INFO=<ID=END,Number=1,Type=String,Description="CNVnator end">'
>> $outfile
echo '##INFO=<ID=CNV_SIZE,Number=1,Type=String,Description="CNVnator
size">' >> $outfile
echo '##INFO=<ID=NORMALIZED_READ_DEPTH,Number=1,Type=String,Description=
"CNVnator normalized read depth">' >> $outfile
echo -e "#CHROM\tPOS\tID\tREF\tALT\tQUAL\tFILTER\tINFO\tFORMAT\tMY_
SAMPLE" >> $outfile

awk 'BEGIN {FS="\t";OFS="\t"} {if (NR>1) {alt="DEL";if($1=="duplication")
{alt="DUP"}; sub("-",":",$2);split($2,a,":");chrom=a[1];pos=a[2];endpo
s=a[3]; print chrom, pos, ":", "N", alt, "1000", "PASS", "END="endpos
";CNV_SIZE="$2 ";NORMALIZED_READ_DEPTH="$4, "GT", "0/1"}}' >> $outfile
```

Now that our CNVnator output is in VCF format, let's annotate it with VEP.

```
infile=my_cnvnator_results.vcf
outfile=my_cnvnator_vep.txt

perl vep/ensembl-vep/vep --tab --format vcf --cache --merged
--everything \
  --offline -i $infile -o $outfile \
  --dir_cache /my/vep/vep_cache --dir_plugins vep/ensembl-vep \
  --use_given_ref --force_overwrite --assembly GRCh38
```

Let's install and run Conanvarvar to identify very large copy number variants. Conanvarvar is available as a Docker image, so we will follow the installation instructions at github.com/VCCRI/ConanVarvar and simply install it with the following:

```
docker pull mgud/conanvarvar:latest
```

Rather than running just one sample, Conanvarvar works better if there is more than one sample. Thus, include more than one BAM input file if possible. The BAM files should appear to be in the one directory. That directory can be soft links to the actual BAM files.

```
mkdir directory_of_bams
ln -s my_sample_1.bam
ln -s my_sample_1.bam.bai
ln -s my_sample_2.bam
ln -s my_sample_2.bam.bai
ln -s my_sample_3.bam
ln -s my_sample_3.bam.bai
```

Now run Conanvarvar.

```
docker run -it --rm -v directory_of_bams:/data mgud/conanvarvar
--bamdir=/data --outdir=/data/output
```

The Conanvarvar output will be in `directory_of_bams/output`. The most reliable way to identify CNVs is to go into the Plots directory and manually look at the plots for each chromosome. A CNV deletion or duplication will be very obvious to see, as shown in the Conanvanvar instructions at `github.com/VCCRI/ConanVarvar`. Most individuals have no true-positive large copy number variants that are more than 1 million base pairs long. Nonetheless, it is good to check this with Conanvarvar, because if there is a very large CNV, then you'd want to know about it, because it probably is having a health impact.

Setting Up AWS Services and Data Storage

Now, let's start the pipeline in AWS. We will create an AWS EFS file storage to hold our data, and an EC2 service instance to perform pipeline run activities on the data.

1. Sign into the AWS Console (see Figure 8.1).

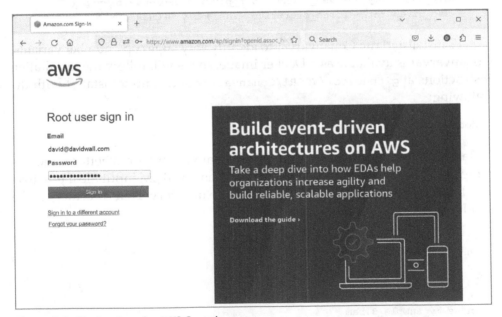

Figure 8.1: Signing into the AWS Console

2. Create an EFS file storage for holding our data. In the upper-left search field, enter **EFS**, and then choose EFS (see Figure 8.2).

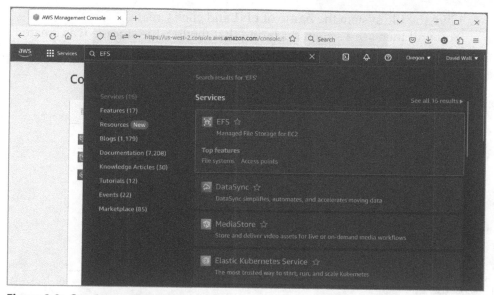

Figure 8.2: Creating an EFS storage volume

3. Click the Create File System button (see Figure 8.3).

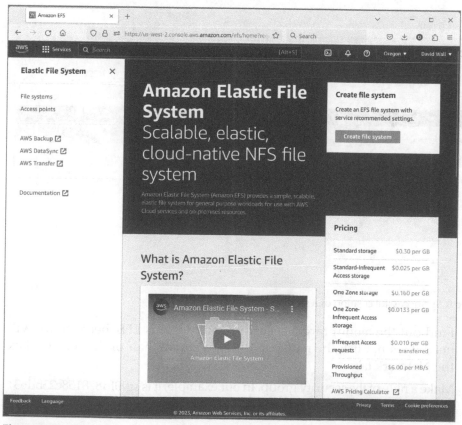

Figure 8.3: Adding a file system

4. Give the file system the name of **efs1** and click Create.
   You will then see a screen declaring that the EFS file storage was created (see Figure 8.4 and Figure 8.5).

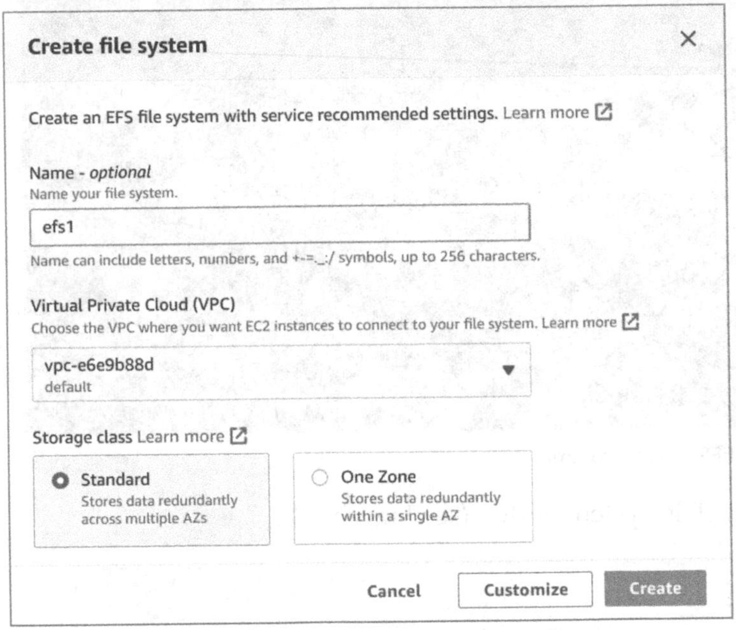

Create file system ✕

Create an EFS file system with service recommended settings. Learn more ⧉

Name - *optional*
Name your file system.

> efs1

Name can include letters, numbers, and +-=._:/ symbols, up to 256 characters.

Virtual Private Cloud (VPC)
Choose the VPC where you want EC2 instances to connect to your file system. Learn more ⧉

> vpc-e6e9b88d
> default ▼

Storage class Learn more ⧉

⦿ **Standard**
Stores data redundantly across multiple AZs

◯ **One Zone**
Stores data redundantly within a single AZ

Cancel   Customize   Create

Figure 8.4: Assigning a name

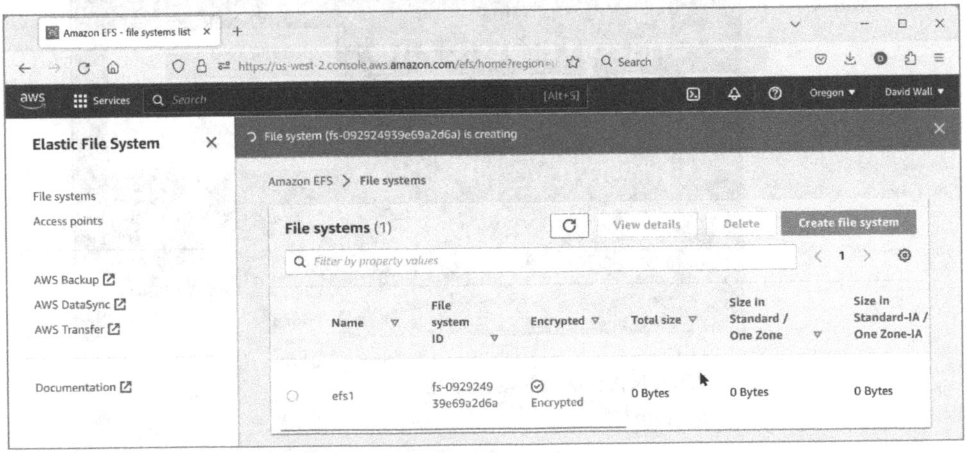

Figure 8.5: Successful creation of a storage volume

5. Find out the mount target security group for this EFS, because we will need it soon. Click the name efs1. In the bottom half of the screen, click the Network tab.

Make a note of the security group. In our example, it is sg-0fd878158c25bd631 (see Figure 8.6).

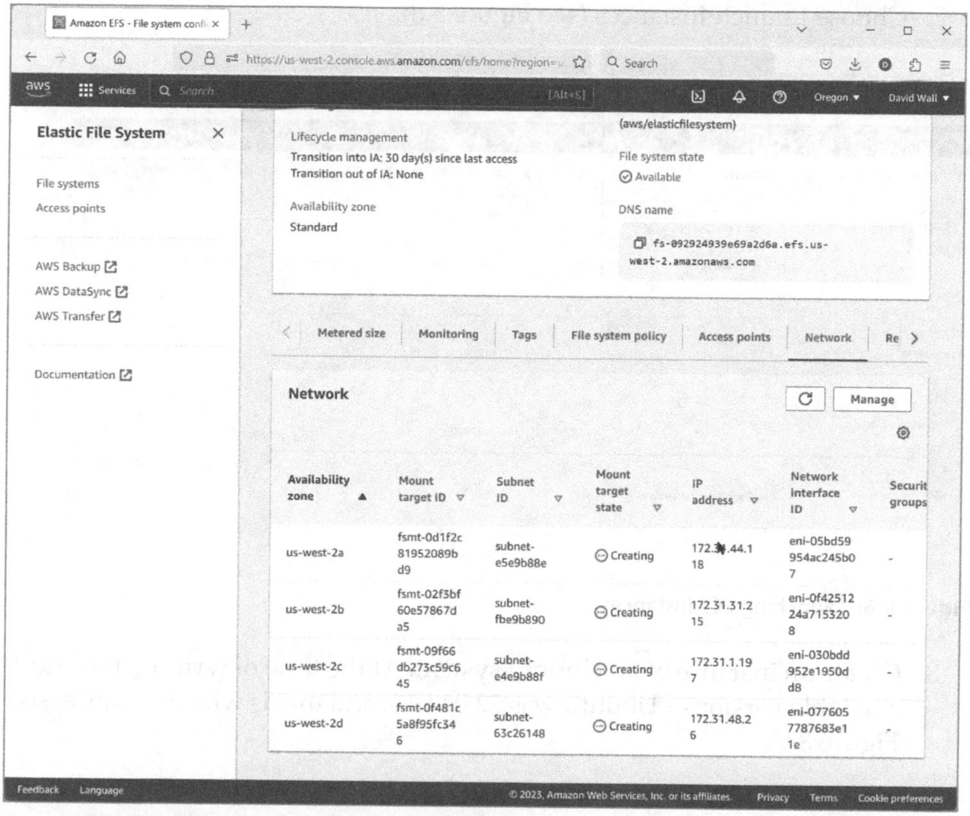

Figure 8.6: Getting the name of the security group

6. Create an instance of an EC2 service to process our data. In the upper-left search field, enter **EC2**, and then choose EC2 (see Figure 8.7).

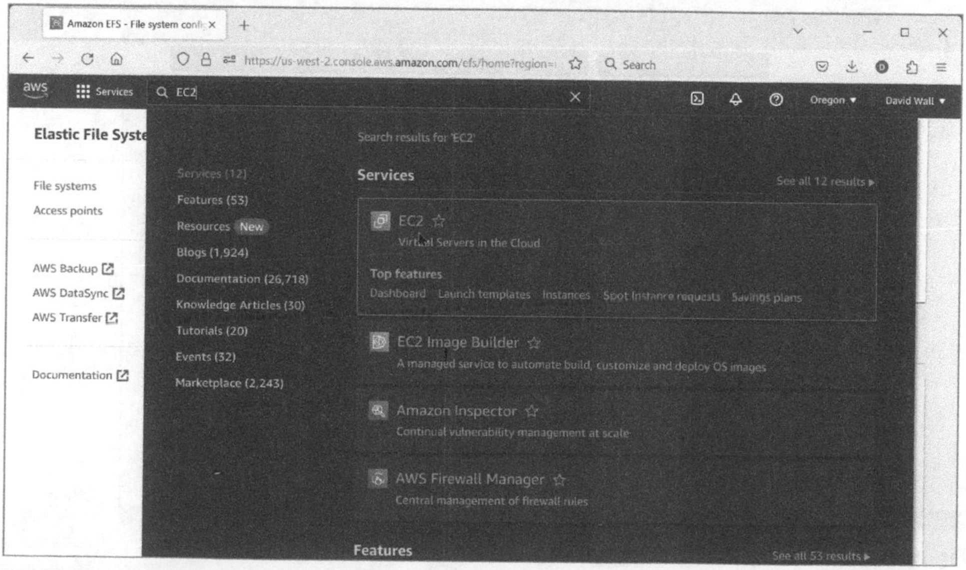

Figure 8.7: The AWS EC2 administration console

7. Choose Launch Instances (see Figure 8.8).

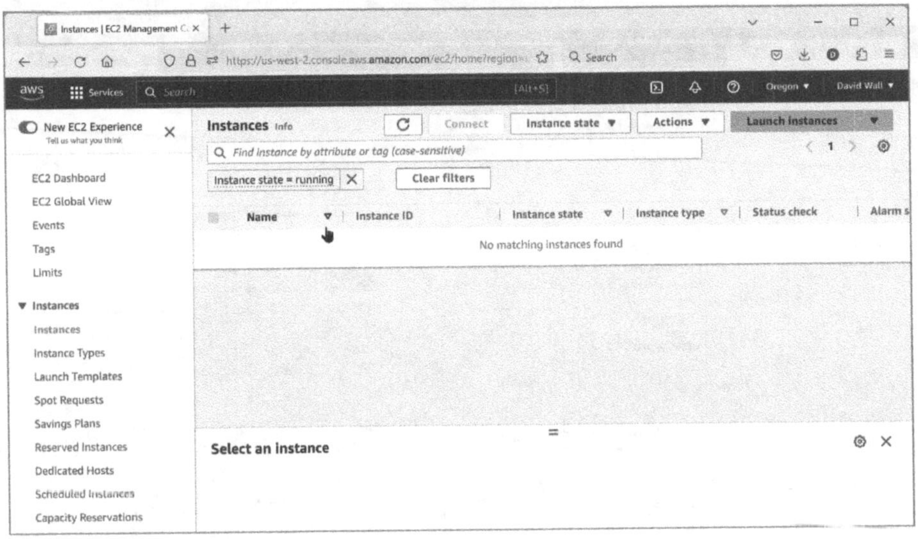

Figure 8.8: Launching EC2 instances

8. Create an instance of an Ubuntu system. At the time of writing, the latest available version of Ubuntu was 22.04 LTS, and that is what we chose (see Figure 8.9).

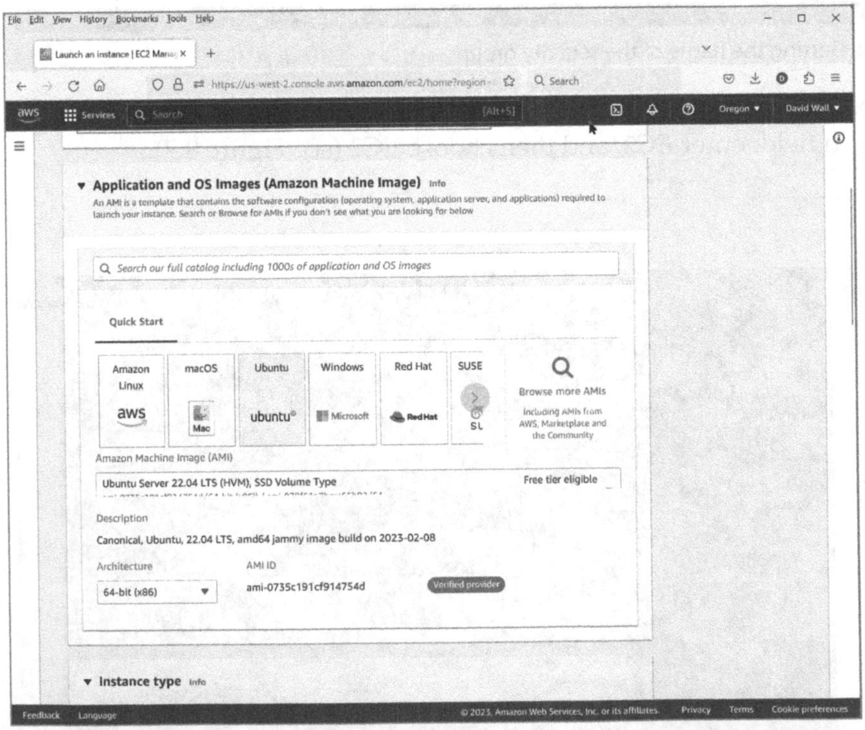

Figure 8.9: Choosing an operating system

9. Choose the t2.micro instance with 1 GB memory and one CPU, and then click Next: Configure Instance Details (see Figure 8.10).

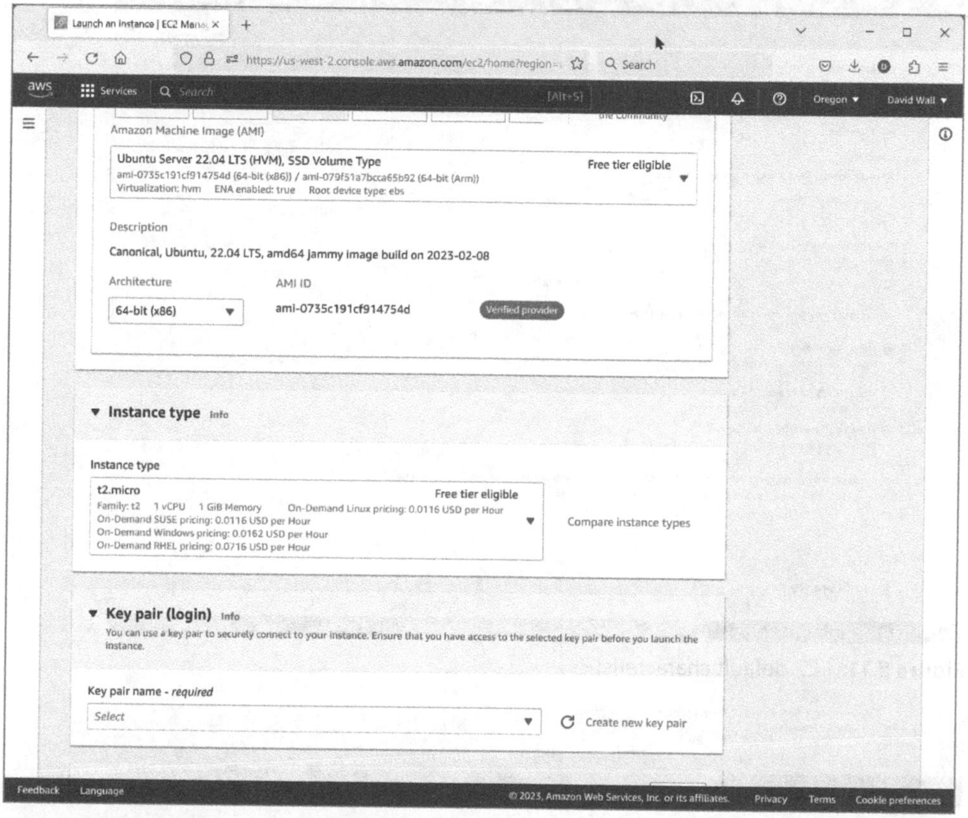

Figure 8.10: Selecting an EC2 instance type

10. Accept the default settings and click Next: Add Storage (see Figure 8.11).

11. Accept the default setting of 8 GB storage and click Next: Add Tags (see Figure 8.12).

12. Continue without creating tags and click Next: Configure Security Group (see Figure 8.13).

13. Keep default security options and click Review And Launch to create the EC2 instance (see Figure 8.14).

14. After reviewing the instance, click Launch to start the launch of the EC2 instance (see Figure 8.15).
The next screen that appears (see Figure 8.16) gives instructions about a key pair, which is a security mechanism for connecting to this EC2 service.

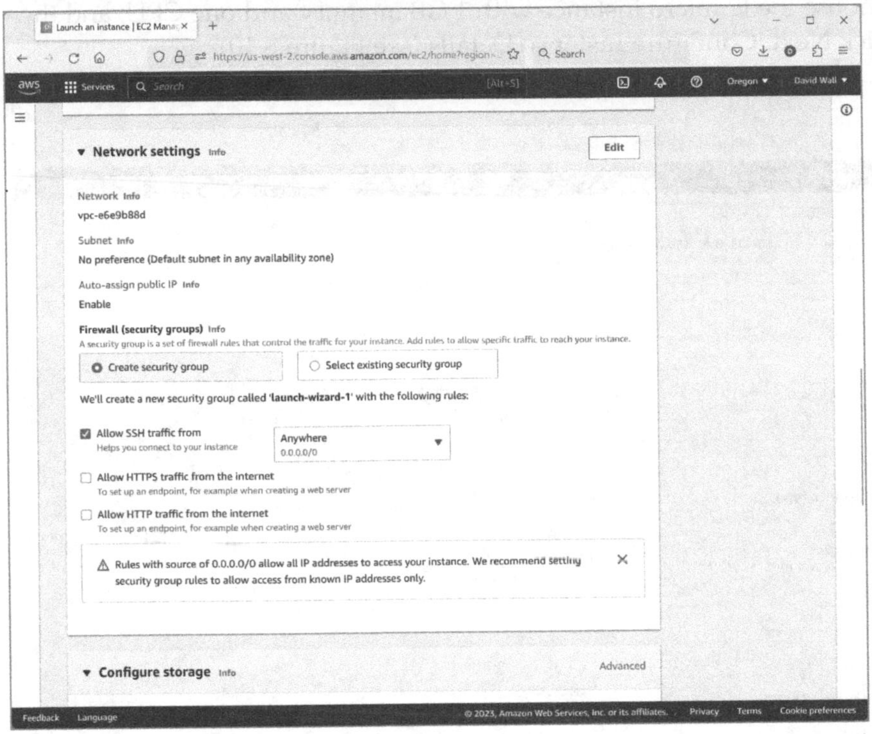

Figure 8.11: EC2 default characteristics

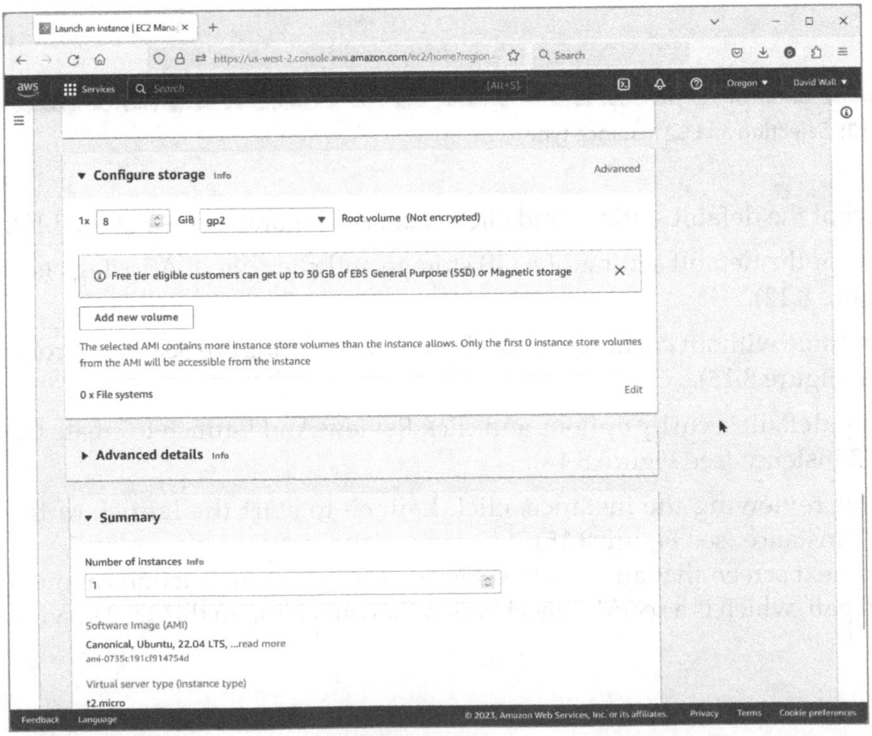

Figure 8.12: Adding storage to an EC2 instance

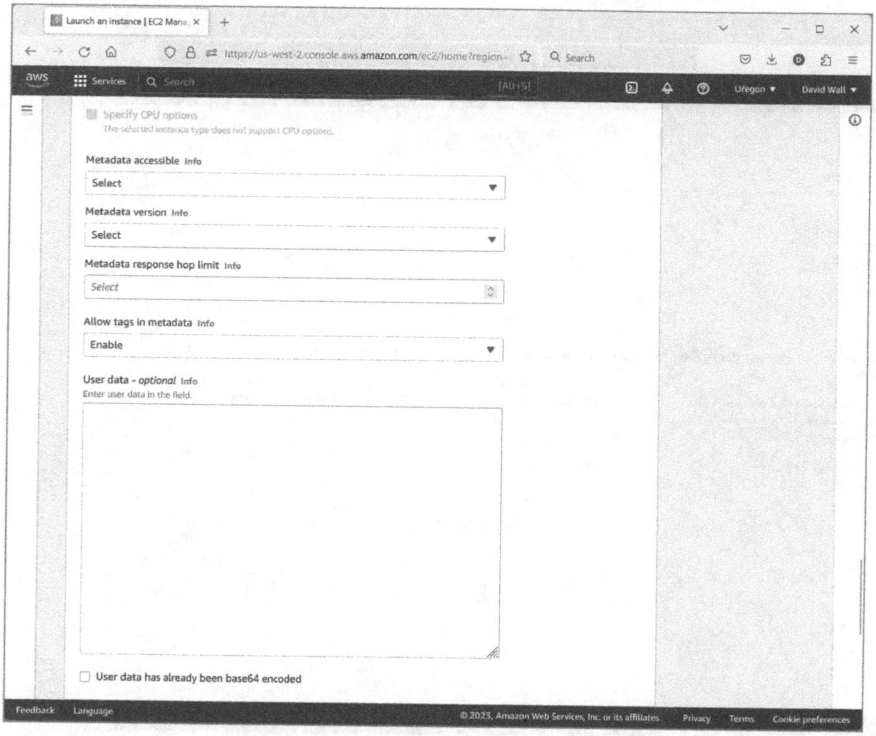

Figure 8.13: Skipping ahead without creating tags

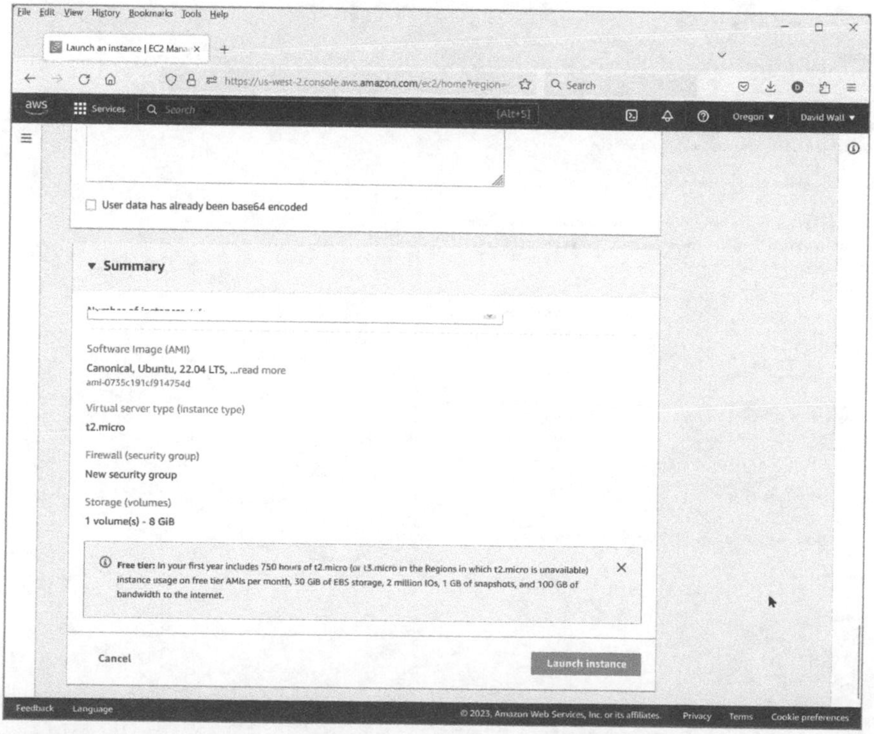

Figure 8.14: Reviewing the characteristics of a new instance

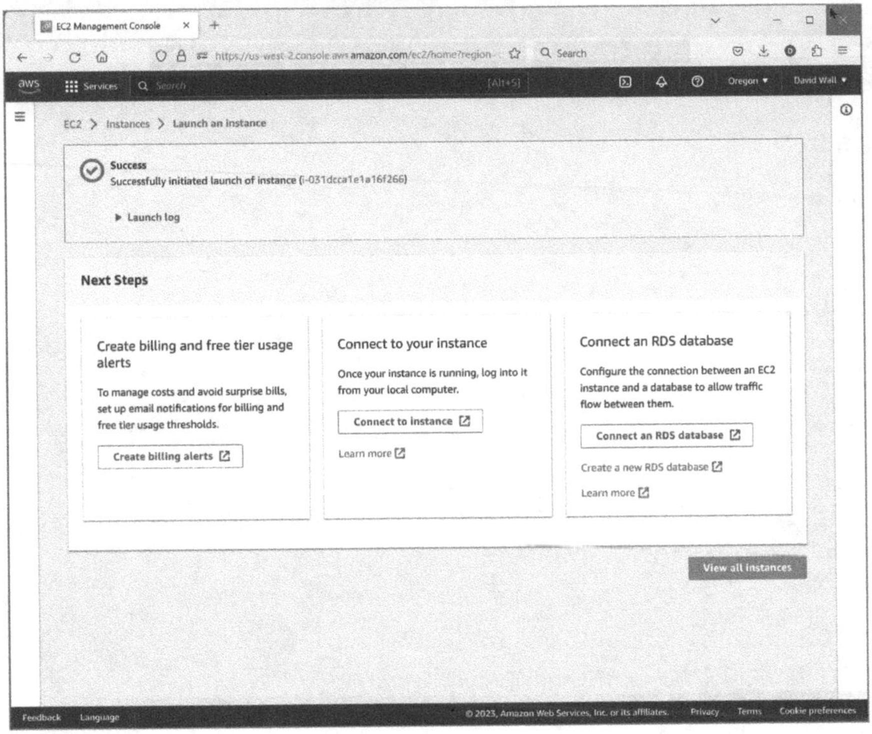

Figure 8.15: Launching a new EC2 instance

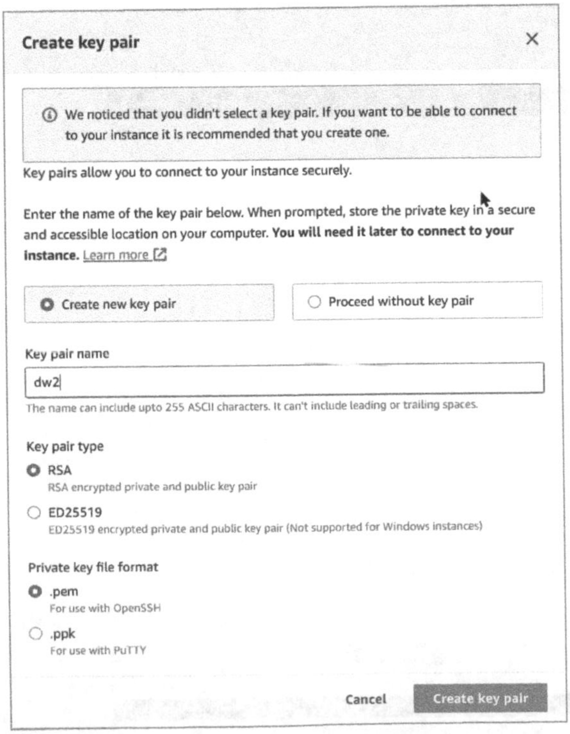

Figure 8.16: Getting the security key pair

15. Choose Create A New Key Pair, fill in a name, and click Download Key Pair to download the PEM file. As instructed, make sure you keep that PEM file for future use. Then you can click Launch Instances to launch the EC2 instance (see Figure 8.17).

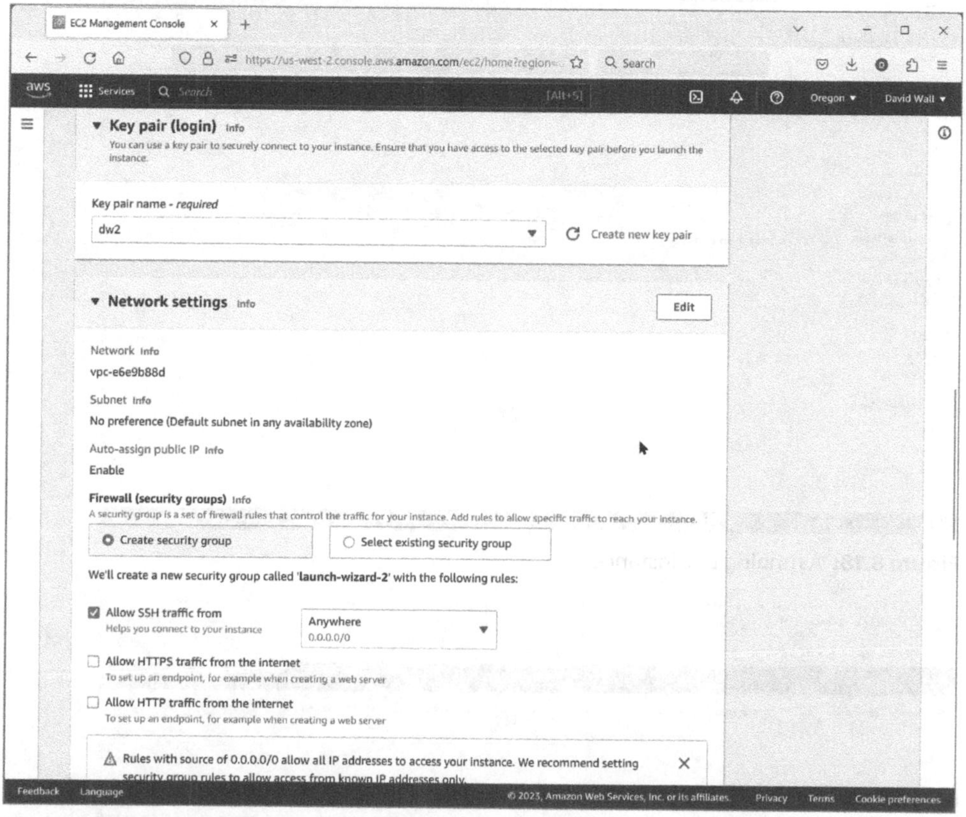

Figure 8.17: Launching a new instance

You will be notified that the EC2 instance is now running (Figure 8.18).

16. If you click View Instances, you will see that the instance has a Running status (See Figures 8.19 and Figure 8.20).

17. When you are not running anything in the EC2 instance, you can stop it. To stop the instance running, select Instance State and then Stop Instance (see Figure 8.21).

18. Confirm by pressing Stop (see Figure 8.22).

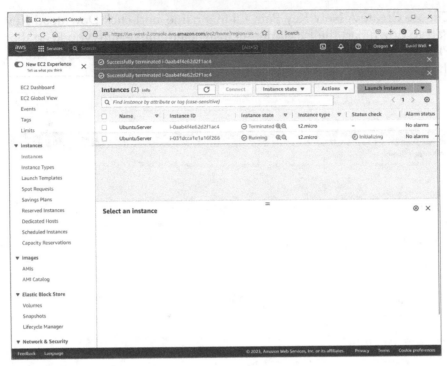

Figure 8.18: A running EC2 instance

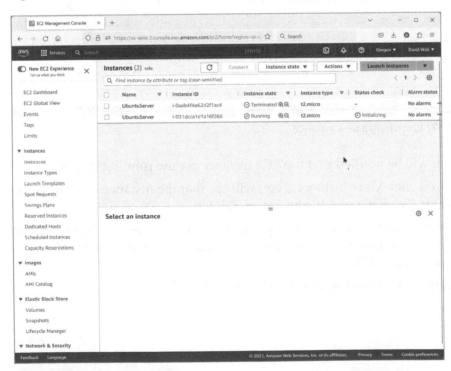

Figure 8.19: Clicking View Instances

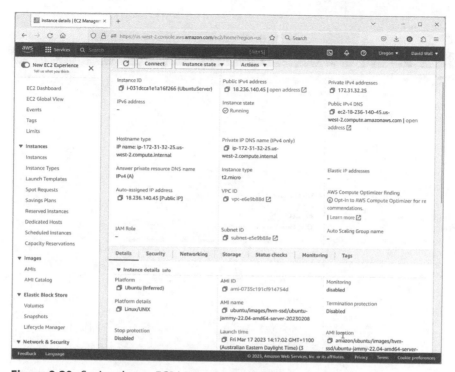

Figure 8.20: Seeing that an EC2 instance is in the Running state

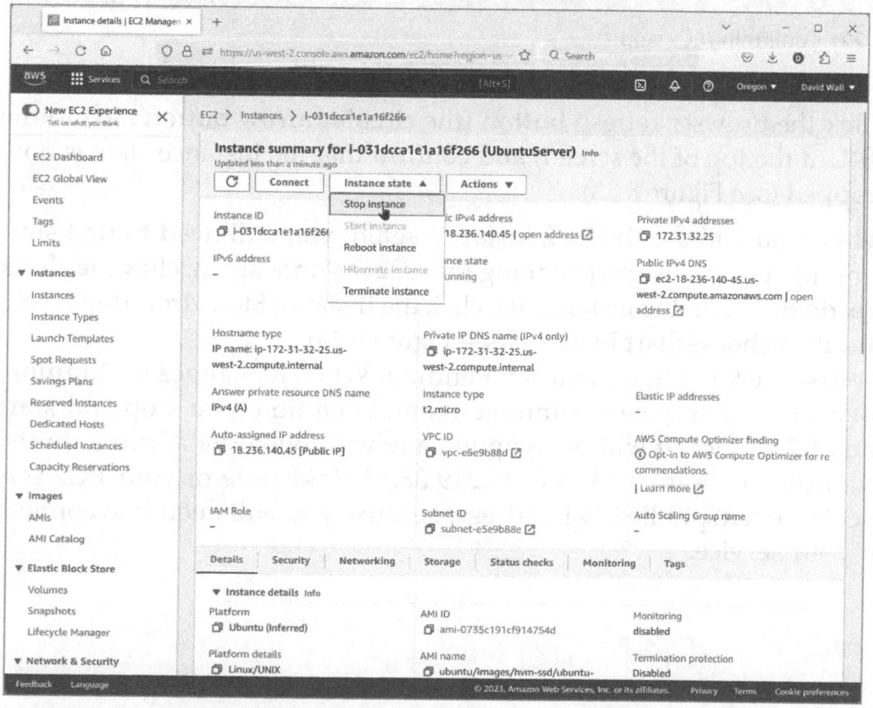

Figure 8.21: Stopping an EC2 instance

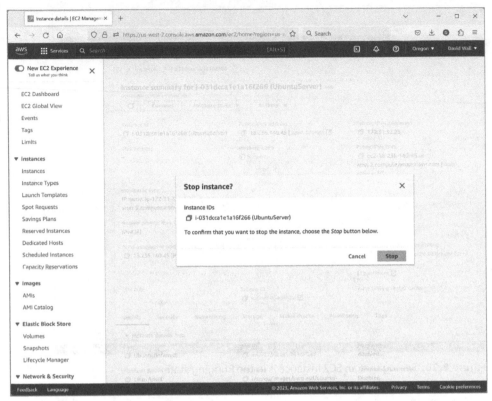

Figure 8.22: Confirming EC2 stop

19. Click the browser refresh button (the circular arrow button next to the URL at the top of the screen) and confirm that the instance state is now Stopped (see Figure 8.23).

20. When you will use the EC2 instance again, you will need to first start running it again. To start running the EC2 instance again, click the check box on the left of the instance ID, click the Instance State drop-down box, and then choose Start Instance (see Figure 8.24).

    At first, Instance State will say Pending. When it changes to Running, then the EC2 service is running again. Each time you stop and start the EC2 service, it will be assigned a new public IPv4 address. In the example in Figure 8.25, it is 34.219.36.57. Take note of your EC2 service's current public IPv4 address, because you will need it to connect to your service.

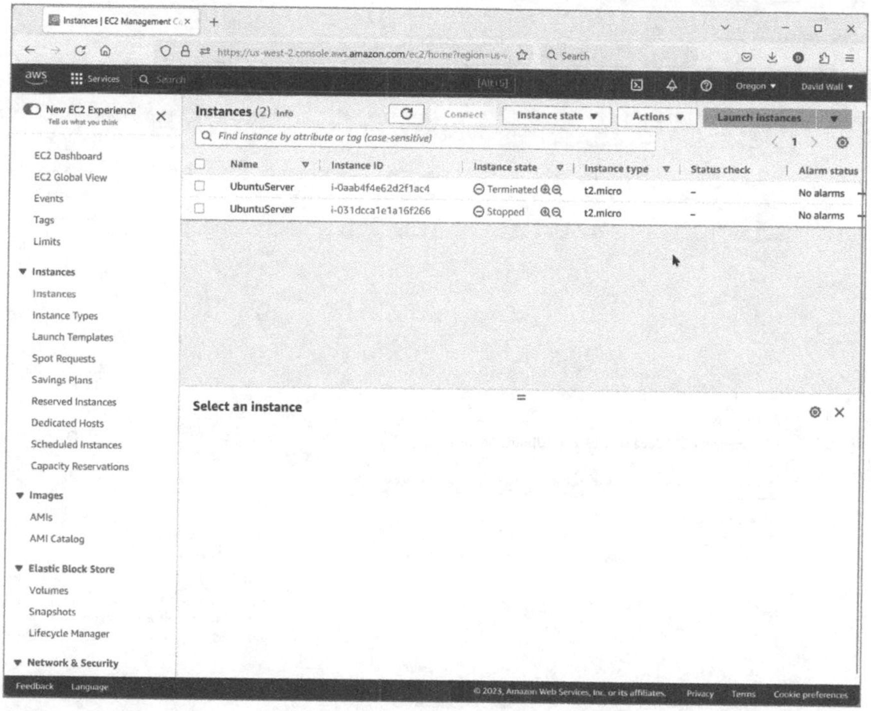

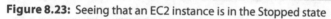

Figure 8.23: Seeing that an EC2 instance is in the Stopped state

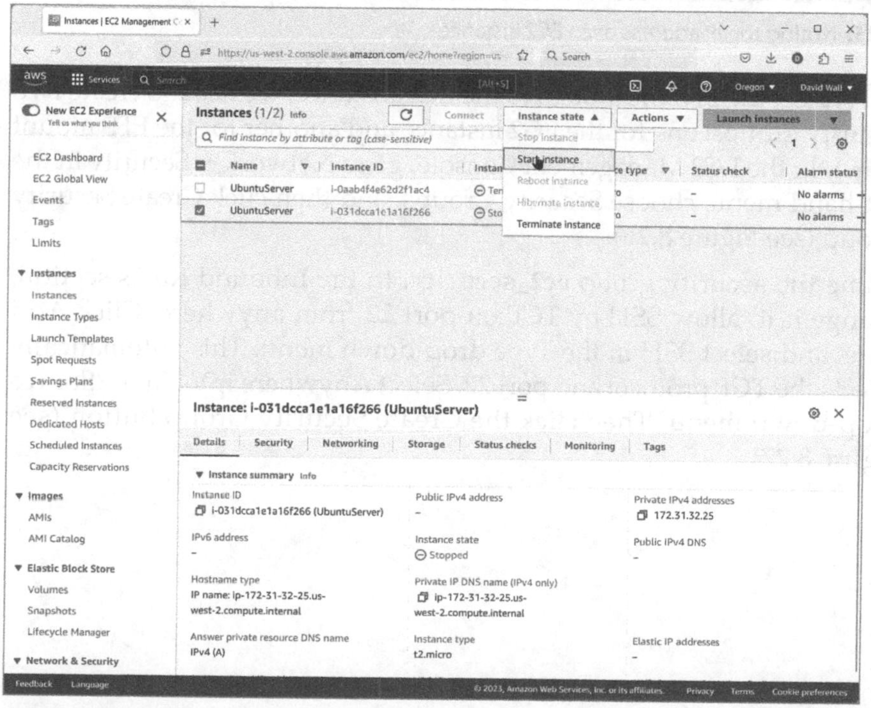

Figure 8.24: Restarting an EC2 instance

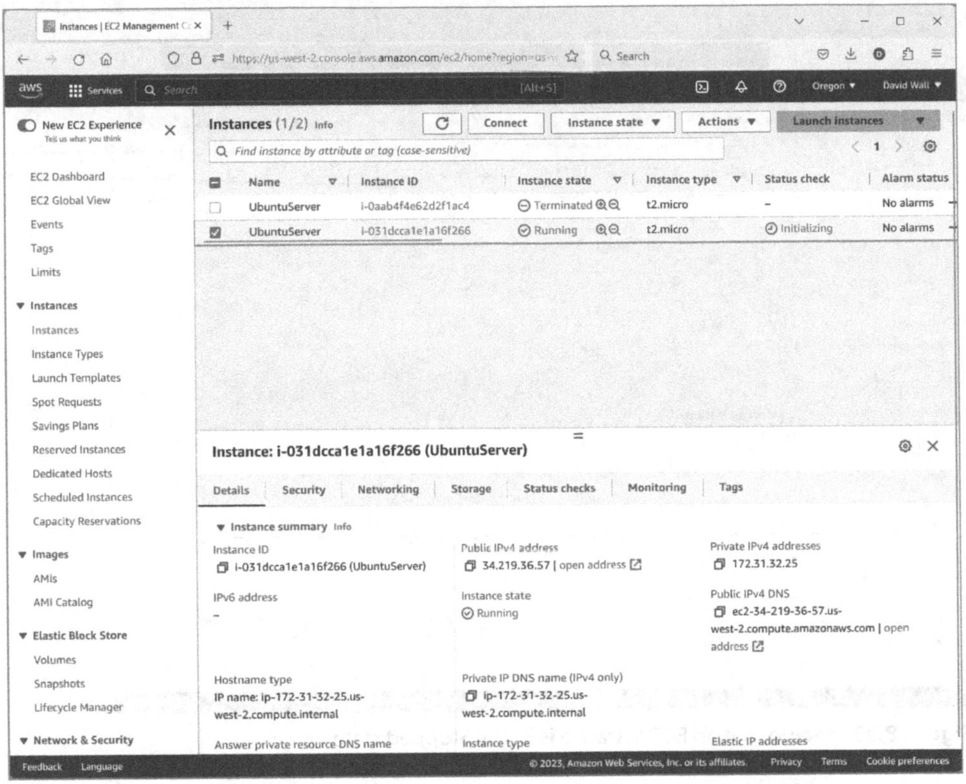

Figure 8.25: Noting the IP address of an EC2 instance

21. To connect the EFS file system to the EC2 instance, we need to create two security groups: one for the EC2 instance and another for the EFS mount target. In the EC2 Management Console, go to Network & Security in the left-hand menu, choose Security Groups, and then click Create Security Group (see Figure 8.26).

22. Name the security group **ec2_security**. In the Inbound Rules section, change it to allow SSH by TCP on port 22 from anywhere. Click Add Rule, and select SSH in the Type drop-down menu. This automatically selects the TCP protocol and port 22. Select Anywhere ipV4 in the Source drop-down menu. Then click the Create Security Group button (see Figure 8.27).

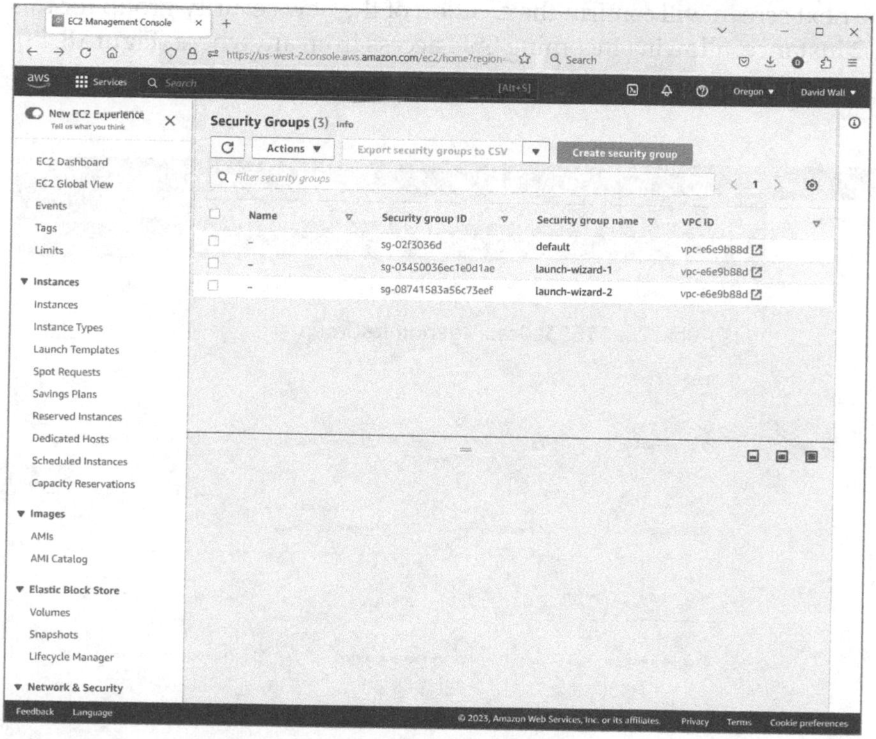

Figure 8.26: Adding a storage volume to an EC2 instance

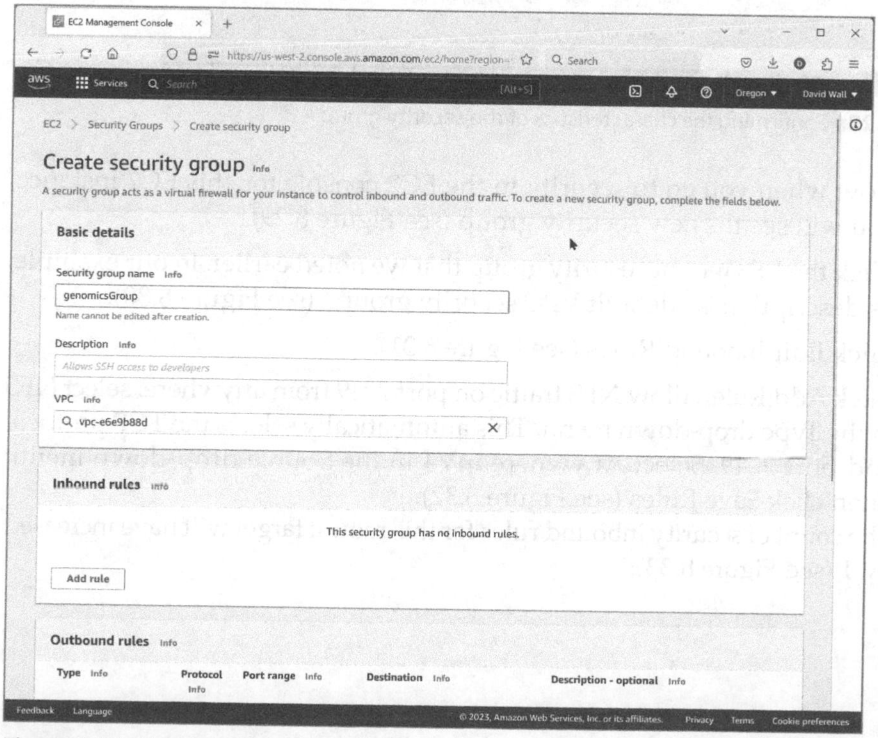

Figure 8.27: Creating a Security Group

The next screen will confirm the creation of the new security group for this EC2 instance, allowing incoming SSH access from anywhere (Figure 8.28).

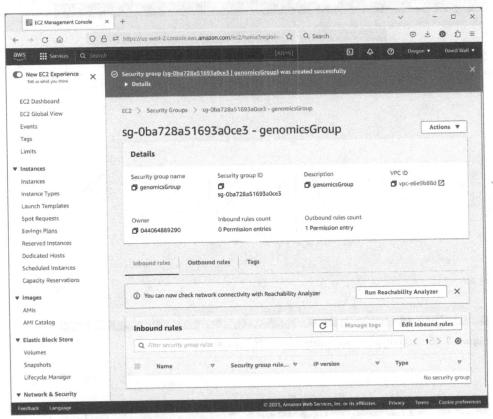

Figure 8.28: Confirming the characteristics of the security group

Now when you go to security in the EC2 console for this EC2 instance, you will see the new security group (see Figure 8.29).

23. Click the EFS mount security group that we noted earlier. In our example, its description is "default VPC security group" (see Figure 8.30).

24. Click Edit Inbound Rules (see Figure 8.31).

25. Click Add Rule. Allow NFS traffic on port 2049 from anywhere: select NFS in the Type drop-down menu. This automatically selects the TCP protocol and port 2049. Select Anywhere ipV4 in the Source drop-down menu. Then click Save Rules (see Figure 8.32).
The count of security inbound rules for this mount target will have increased by 1 (see Figure 8.33).

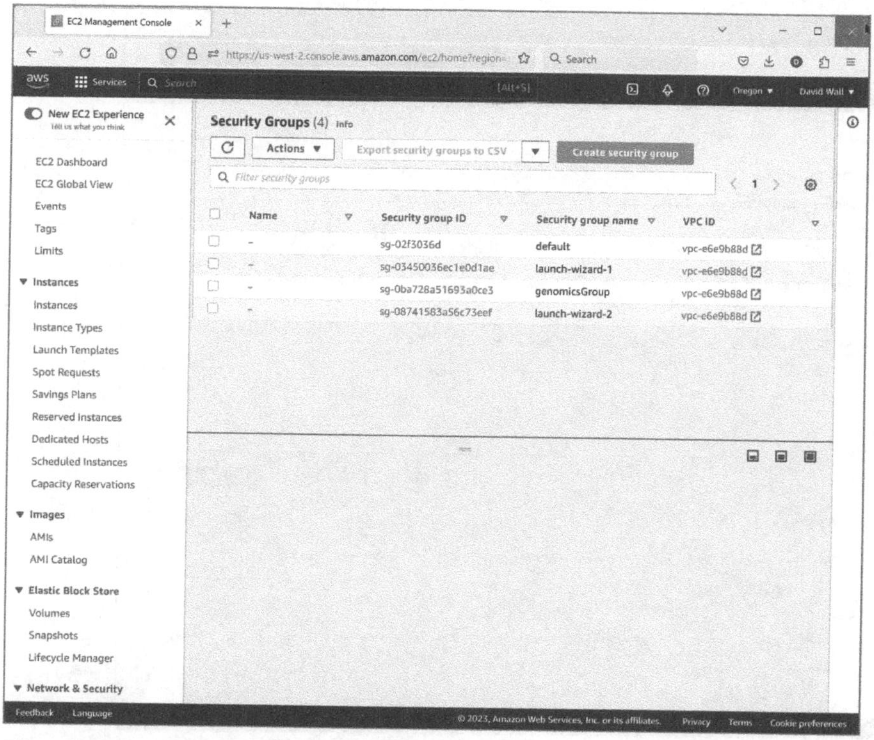

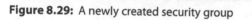

Figure 8.29: A newly created security group

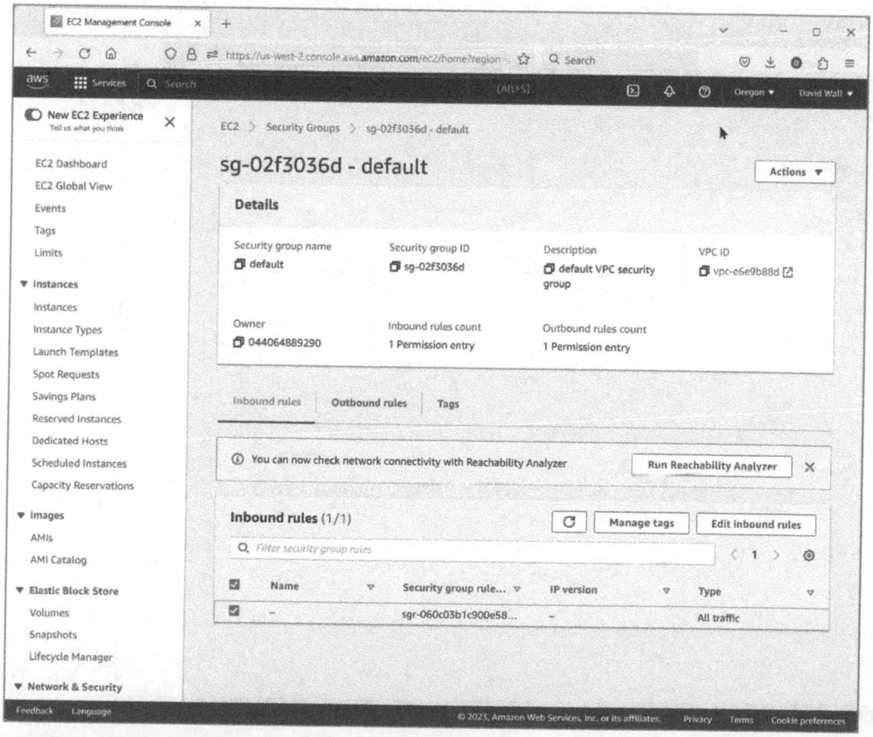

Figure 8.30: Selecting the EFS mount security group

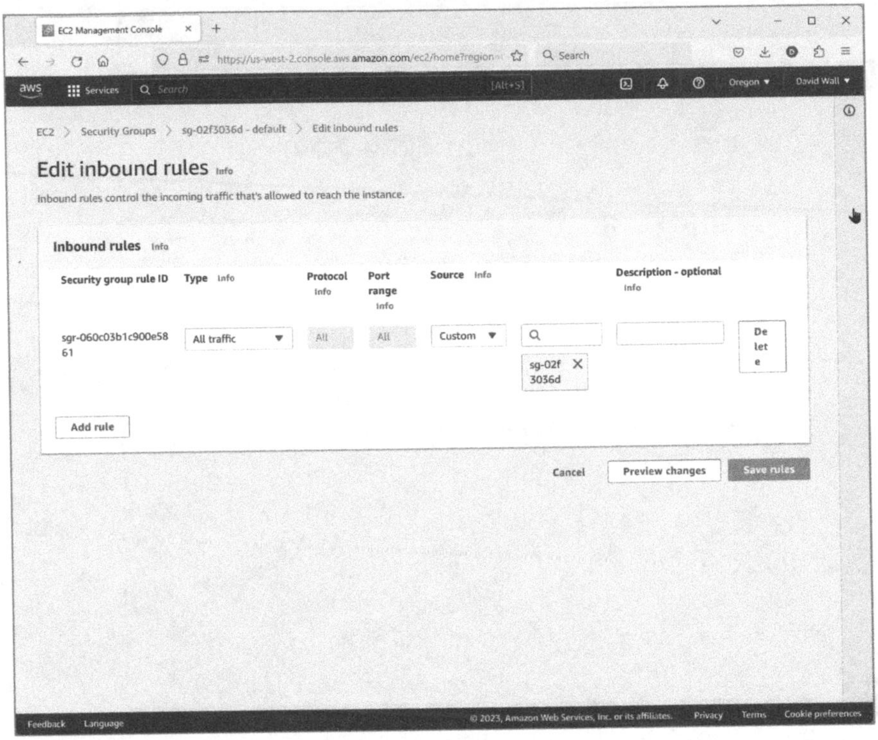

Figure 8.31: Editing inbound security group rules

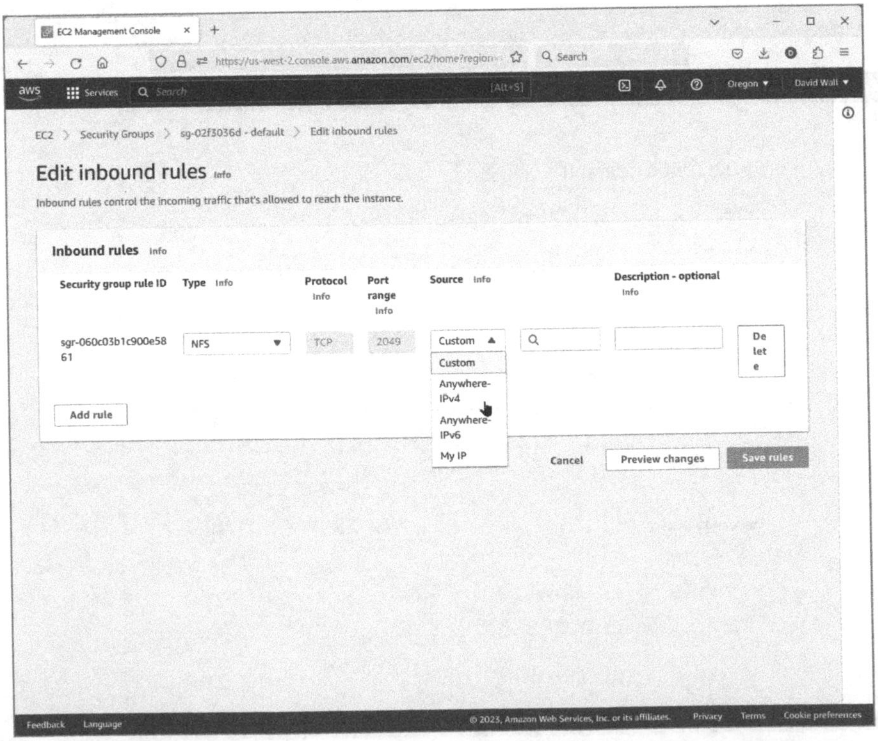

Figure 8.32: Adding a rule

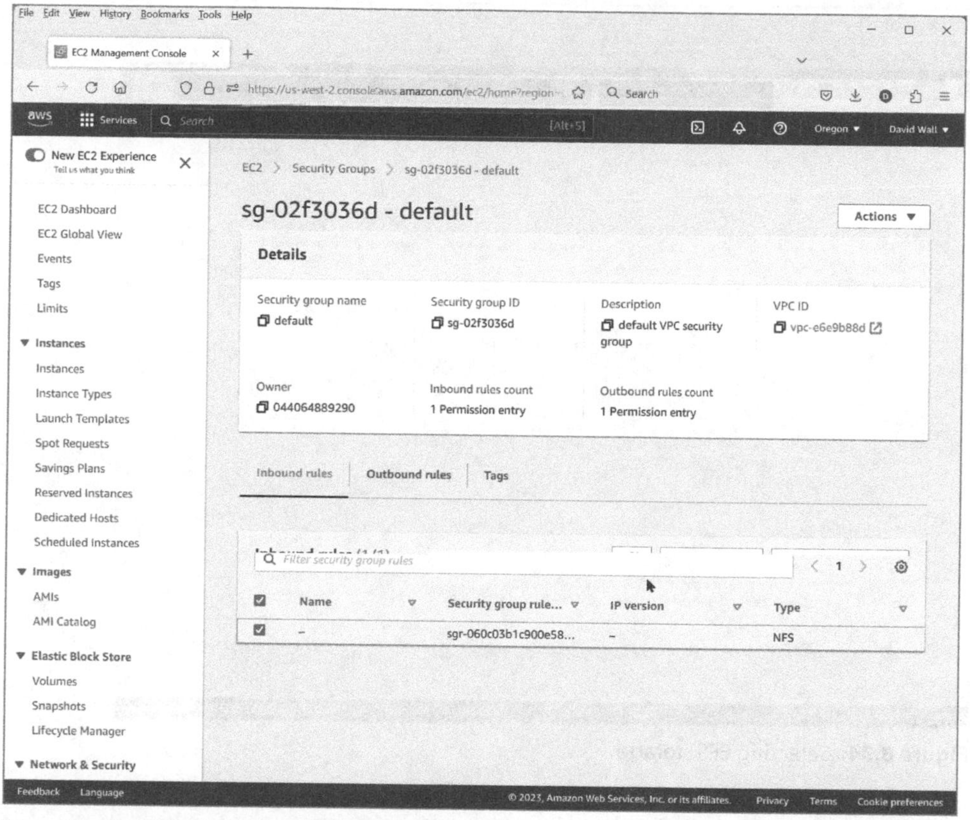

Figure 8.33: Noting the creation of the new rule

26. Next we connect our EFS file storage to our EC2 service instance. Go to the EFS file storage (type **EFS** in the search box at the top of the screen) and click efs1 (see Figure 8.34).

27. Click the Attach button (see Figure 8.35).
    It will display mount information. Take a note of the information (see Figure 8.36):

```
sudo mount -t efs -o tls fs-08945546ab24ce0a5:/ efs
sudo mount -t nfs4 -o nfsvers=4.1,rsize=1048576,wsize=1048576,hard,
timeo=600,retrans=2,noresvport fs-08945546ab24ce0a5.efs.us-east-1.
amazonaws.com:/ efs
```

Next we will log in to our EC2 service to attach it to the EFS data storage using the preceding mount command.

28. If you are using a Windows system, the puTTY software can be used to connect to the AWS EC2 service instance. First run the puTTYgen software to convert the AWS PEM file to a PPK file that puTTY needs. Ensure that Type Of Key To Generate is RSA, click Load, choose the PEM file, and then click Save Private Key. Give it the same file name as the PEM file, with a PPK suffix.

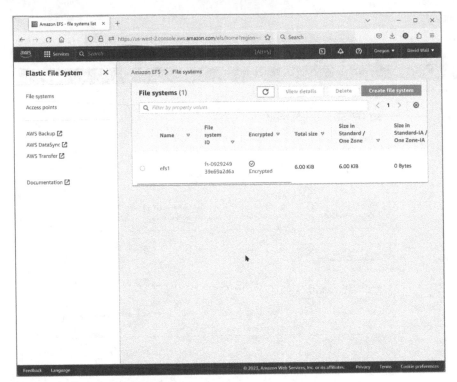

Figure 8.34: Selecting EFS storage

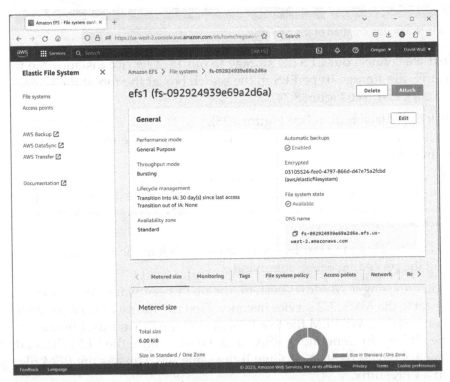

Figure 8.35: Attaching an EFS volume

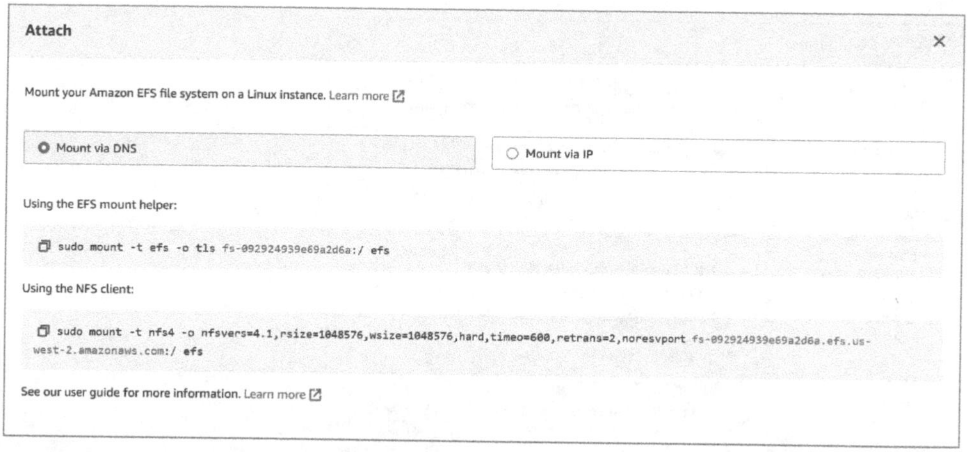

Figure 8.36: Information about the mounted volume

29. Start puTTY. Put **ubuntu** plus the @ symbol and the current Public IPv4 address of the EC2 instance in the Host Name (or IP Address) field of the puTTY screen. In our example, it is ubuntu@44.203.117.9 (see Figure 8.37).

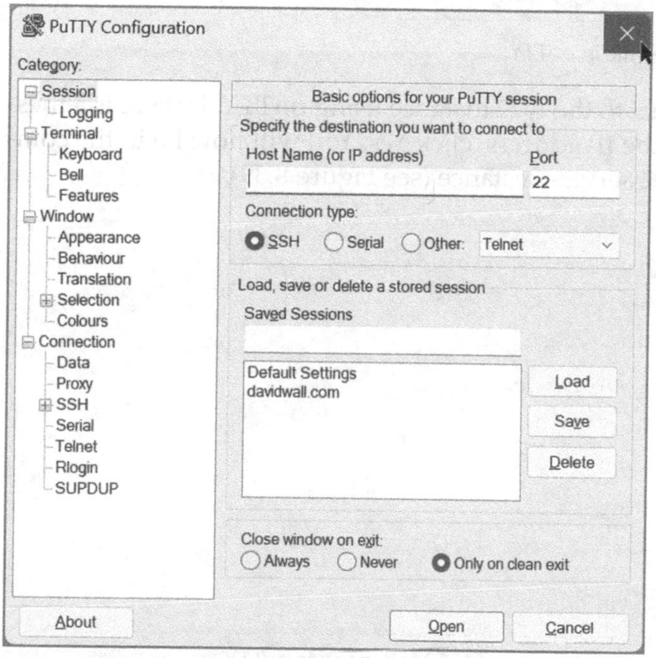

Figure 8.37: Specifying a host in puTTY

30. In the Category panel, choose Connection, then SSH, and then Auth. Click Browse and choose the PPK file you just created from the PEM file (see Figure 8.38).

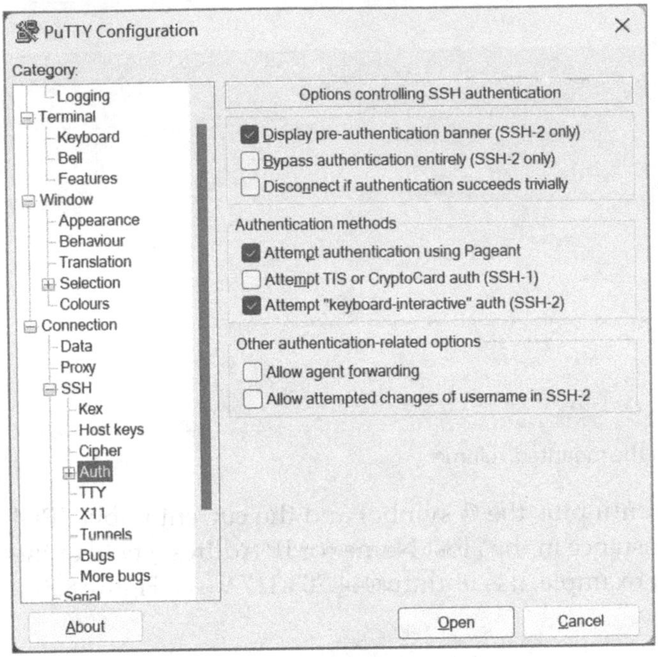

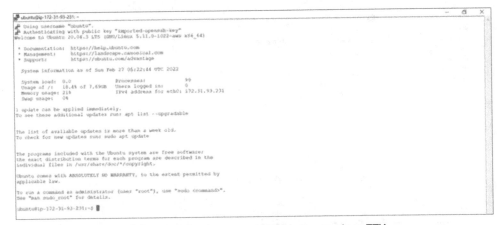

Figure 8.38: Selecting a PPK key file in puTTY

31. Click Open to connect to the EC2 instance using puTTY. If there is a message about caching the IP address, click Yes. You will now be in the command line of the EC2 service instance (see Figure 8.39).

Figure 8.39: A successful connection to a remote EC2 instance, via puTTY

32. Run the following command to create a directory for our EFS:

```
sudo mkdir efs
```

33. Install the software necessary for accessing the EFS:

```
sudo apt-get -y install nfs-common
```

Messages will appear on the screen about the progress of the installation, and when it is done, the cursor will return to the $ prompt.

34. Run the `mount` command from step 27. Every time you stop and restart the EC2 service, you will need to rerun this command.

```
sudo mount -t nfs4 -o nfsvers=4.1,rsize=1048576,wsize=1048576,hard,
timeo=600,retrans=2,noresvport fs-08945546ab24ce0a5.efs.us-east-1.
amazonaws.com:/ efs
```

If there is a problem with the security rules that we set up previously, then you will receive an error message.

```
mount.nfs4: Connection timed out
```

If the EFS mount in EC2 works, then you will simply receive the Linux $ prompt.

35. Confirm that the EC2 can now see and access the EFS. The command:

```
df -h
```

returns the following:

```
fs-08945546ab24ce0a5.efs.us-east-1.amazonaws.com:/  8.0E     0  8.0E
0% /home/ubuntu/efs
```

The command:

```
ls -l
```

returns the following:

```
drwxr-xr-x 2 root root 6144 Mar  5 02:58 efs
```

36. Make sure that we can write to the EFS by trying to edit a file on the EFS:

```
nano efs/test.txt
```

Add some text to the file:

```
this is a test
```

Press Ctrl+X to finish entering text, press Ctrl+Y when asked to save the buffer, and then press Enter again to save to the filename `efs/test.txt`.

If you are not able to write or save a file to the EFS, then change the permissions to allow writing.

```
sudo chmod go+w efs/
```

37. Create the `efs/test.txt` file as a test of writing. Confirm that you can read what you have written to the EFS.

```
cat efs/test.txt
```

It should return what you typed and saved.

```
this is a test
```

From our EC2 instance, we can now successfully access our EFS file storage.

Copying the FASTQ Files

FASTQ files contain the sequencing data from the sequencing service provider. For this exercise, we will obtain the publicly accessible whole-genome sequencing FASTQ files of an individual. The individual is sample NA12878 from the 1000 Genomes project, also known as *Utah Woman*. The sequencing data from this individual is now a standard, used by sequencing centers and bioinformaticians around the world for quality control and benchmarking. If you have your own FASTQ files of sequencing paired reads that you want to process, then you can use them instead of obtaining the data for NA12878.

Use puTTY to ssh to your EC2, mount the EFS, and type the following commands:

```
cd efs
mkdir -p data/fastq
cd data/fastq
wget ftp://ftp.sra.ebi.ac.uk/vol1/fastq/SRR622/SRR622457/SRR622457_1.
fastq.gz
wget ftp://ftp.sra.ebi.ac.uk/vol1/fastq/SRR622/SRR622457/SRR622457_2.
fastq.gz
```

These `wget` commands will obtain the paired sequence read files for NA12878 whole-genome sequencing. The downloads will take hours because the files are large. During the download, you will see reporting of the progress of the downloading.

```
SRR622457_1.fastq.gz 0%[                    ]           0  --.-KB/s
SRR622457_1.fastq.gz 1%[>                   ] 794.44M  8.62MB/s    eta 1h 55m
```

You can open a second puTTY terminal and log in to the instance a second time and download the second FASTQ file there while downloading the first FASTQ using the first terminal. When the downloading is finished after a few hours, you can see the size of the files.

```
ls -lh
shows that the first fastq file is 48 GB and the second is 67 GB in size.
-rw-rw-r-- 1 ubuntu ubuntu 48G Mar  5 17:36 SRR622457_1.fastq.gz
-rw-rw-r-- 1 ubuntu ubuntu 67G Mar  5 17:22 SRR622457_2.fastq.gz
```

If you look in the EFS console (see Figure 8.40), you will see that the EFS contains nearly 115 GB of data.

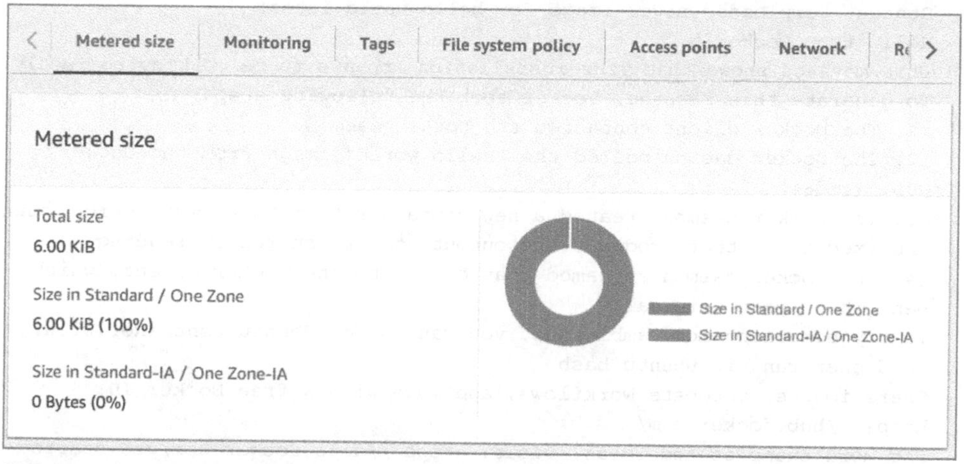

Figure 8.40: Observing the contents of an EFS volume

Installing Docker and Containers

Some of the genomics pipeline software that we want to run is already available in Docker containers. That means that instead of having to install the software and their dependencies as shown previously in this chapter, we can simply use existing Docker containers that contain the software already installed.

Let's install Docker so that we can use Docker containers. Inside your running EC2 service, type the following commands. When asked whether you want to install extra packages, enter Y for Yes.

```
sudo apt update
sudo apt install docker.io
sudo chmod 666 /var/run/docker.sock
```

Make sure Docker is installed. The following command

```
docker --version
```

should display something similar to the following:

```
Docker version 20.10.7, build 20.10.7-0ubuntu5~20.04.2
```

And running the following

```
docker run hello-world
```

should display several lines of messages, including these:

```
Unable to find image 'hello-world:latest' locally
latest: Pulling from library/hello-world
2db29710123e: Pull complete
Digest: sha256:97a379f4f88575512824f3b352bc03cd75e239179eea0fecc38e59
7b2209f49a
```

```
Status: Downloaded newer image for hello-world:latest
Hello from Docker!
This message shows that your installation appears to be working correctly.
To generate this message, Docker took the following steps:
 1. The Docker client contacted the Docker daemon.
 2. The Docker daemon pulled the "hello-world" image from the Docker
Hub. (amd64)
 3. The Docker daemon created a new container from that image which runs
the executable that produces the output you are currently reading.
 4. The Docker daemon streamed that output to the Docker client, which
sent it to your terminal.
To try something more ambitious, you can run an Ubuntu container with:
 $ docker run -it ubuntu bash
Share images, automate workflows, and more with a free Docker ID:
https://hub.docker.com/
For more examples and ideas, visit: https://docs.docker.com/get-started/
```

Now that Docker is installed, let's obtain the container from Broad Institute that contains a lot of the programs we want to run in our bioinformatics pipeline.

```
docker pull broadinstitute/genomes-in-the-cloud:2.3.1-1512499786
```

This will produce lines, such as the following, to show that the container is being retrieved to the EC2:

```
2.3.1-1512499786: Pulling from broadinstitute/genomes-in-the-cloud
5040bd298390: Pull complete
fce5728aad85: Pull complete
c42794440453: Pull complete
0c0da797ba48: Pull complete
7c9b17433752: Pull complete
114e02586e63: Pull complete
e4c663802e9a: Pull complete
7904482546ea: Pull complete
57109deba517: Pull complete
9d04d94be61e: Pull complete
4056023af14c: Pull complete
5c1001b594f3: Pull complete
35e028448590: Pull complete
Digest: sha256:4fca8ca945c17fd86e31eeef1c02983e091d4f2cb437199e74b16
4d177d5b2d1
Status: Downloaded newer image for broadinstitute/
genomes-in-the-cloud:2.3.1-1512499786
docker.io/broadinstitute/genomes-in-the-cloud:2.3.1-1512499786
```

You can test that the Docker pull has worked with the following command:

```
docker run -it --rm  broadinstitute/genomes-in-the-cloud:2.3.1-1512499786
```

This will deposit you into the container, where the command

```
ls -l
```

will show you software available in the container:

```
-rw-r--r--  1 root root      2501 Dec   5   2017 Dockerfile
-rwxr-xr-x  1 root root  12763805 Dec   5   2017 GATK34.jar
-rw-r--r--  1 root root  13182715 Dec   5   2017 GATK35.jar
-rw-r--r--  1 root root  14205988 Dec   5   2017 GATK36.jar
-rwxr-xr-x  1 root root   1793869 Dec   5   2017 bgzip
-rwxr-xr-x  1 5494 1015    345712 May  31   2016 bwa
drwxr-xr-x  2 root root      4096 Dec   5   2017 gatk4
-rw-r--r--  1 root root  12370448 Dec   5   2017 picard.jar
drwxr-xr-x 11 root root      4096 Dec   5   2017 svtoolkit2.00
-rwxr-xr-x  1 root root    217146 Dec   5   2017 tabix
```

To exit the container, type the following:

```
exit
```

The data that is outside the container needs to be bound to the Docker container so that the programs located inside the container can access the data that is outside the container.

The /home/ubuntu/efs/data:/home/data part of the docker run command binds /home/ubuntu/efs/data, which is outside the container, to /home/data inside the container.

```
docker run -v /home/ubuntu/efs/data:/home/data -it --rm  broadinstitute/
genomes-in-the-cloud:2.3.1-1512499786
```

This brings you inside the Docker container, as shown by this prompt:

```
root@f53ed17480fe:/usr/gitc#
```

While we are inside the container, we can look in /home/data, which is accessing the /home/ubuntu/efs/data data outside the container.

```
ls -l /home/data/fastq
```

The result of this command lists the files that we put on our EFS.

```
-rw-rw-r-- 1 1000 1000 50540242629 Mar  5 17:36 SRR622457_1.fastq.gz
-rw-rw-r-- 1 1000 1000 71362146280 Mar  5 17:22 SRR622457_2.fastq.gz
```

While inside the container, we can create more data on the EFS.

```
echo 'this is a test' > /home/data/test.txt
```

When we exit the container by typing `quit`, we can see the data we created on the EFS.

```
cat /home/ubuntu/efs/data/test.txt
```

This command will display the following:

```
this is a test
```

We want to download more Docker containers, but there will not be enough room in the default root directory for Docker. So let's put our Docker containers onto our EFS.

You can see what Docker images you have with this command:

```
docker images -a
```

That will show the image ID.

```
REPOSITORY                          TAG               IMAGE ID     CREATED
SIZE
broadinstitute/genomes-in-the-cloud 2.3.1-1512499786  7b3b42669b20 4 years ago
1.88GB
```

Use that image ID to remove the Docker image.

```
docker rmi 7b3b42669b20
```

Various "Untagged" and "Deleted" messages will let you know that it is removing the image.

Let's move the Docker root directory to our EFS. Stop Docker from running while we make the change.

```
sudo systemctl stop docker.service
sudo systemctl stop docker.socket
```

Edit the Docker configuration file:.

```
sudo nano /lib/systemd/system/docker.service
```

Find this line in it:

```
ExecStart=/usr/bin/dockerd -H fd:// --containerd=/run/containerd/
containerd.sock
```

And change it to this:

```
ExecStart=/usr/bin/dockerd -g /home/ubuntu/efs/docker -H fd:// --
containerd=/run/containerd/containerd.sock
```

Press Ctrl+X to quit the nano editor, entering Y for Yes to save the changes to the configuration file. Find the existing Docker containers. Create the new directory for Docker images.

```
sudo mkdir -p /home/ubuntu/efs/docker
```

Start Docker again, pointing at the new directory.

```
sudo systemctl daemon-reload
sudo systemctl start docker
```

Make sure it is pointing at the new directory.

```
ps aux | grep -i docker | grep -v grep
```

This command should display the following:

```
root        11212  6.8  9.1 840764 90656 ?         Ssl  06:30   0:33 /
usr/bin/dockerd -g /home/ubuntu/efs/docker -H fd:// --containerd=/run/
containerd/containerd.sock
```

Get the Docker image again from Broad Institute, and store it this time on the EFS in /home/ubuntu/efs/docker, as specified in /lib/system/system/docker.service.

```
docker pull broadinstitute/genomes-in-the-cloud:2.3.1-1512499786
```

Test that this new image works.

```
docker run -v /home/ubuntu/efs/data:/home/data -it broadinstitute/
genomes-in-the-cloud:2.3.1-1512499786
```

This should again bring you to the root@4016dc656301:/usr/gitc# inside the container. Again, exit the container by typing exit.

If you have stopped and restarted your EC2 instance since the last time you used this Docker container, then you may see the following error message:

```
Got permission denied while trying to connect to the Docker daemon
socket at unix:///var/run/docker.sock: Post http://%2Fvar%2Frun%2F
docker.sock/v1.24/images/create?fromImage=broadinstitute%2Fgenomes-in-
the-cloud&tag=2.3.1-1512499786: dial unix /var/run/docker.sock: connect:
permission denied
#####
```

Fix this problem with this command:

```
sudo chmod 666 /var/run/docker.sock
```

The next problem may be this one:

```
failed to register layer: symlink
../16fd4589a8f473050b2d35db63d1c2b5e4cb54bb941b5a6a298138c1447276db/diff
/home/ubuntu/efs/docker/overlay2/l/TPABFM3YD63Q5WT5E2SB7BNG67: no such
file or directory
```

If you have that problem, then fix it with the following:

```
sudo service docker stop
sudo service docker start
```

You should now be able to go into this `broadinstitute/genomes-in-the-cloud:2.3.1-1512499786` container with the `docker run` command.

Now that we have enough room to retrieve more Docker containers, let's download the Docker container containing GATK resources.

```
docker pull us.gcr.io/broad-gatk/gatk:4.3.0.0
```

The successful execution of that next `docker pull` command will display the following:

```
4.3.0.0: Pulling from broad-gatk/gatk
1c37c9723c25: Pull complete
Digest: sha256:b3bde7bc74ab00ddce342bd511a9797007aaf3d22b9cfd7b52f41
6c893c3774c
Status: Downloaded newer image for us.gcr.io/broad-gatk/gatk:4.3.0.0
us.gcr.io/broad-gatk/gatk:4.3.0.0
```

Let's take a look inside that new container.

```
docker run -v /home/ubuntu/efs/data:/home/data -it us.gcr.io/broad-gatk/
gatk:4.3.0.0
```

Now let's follow the pipeline jobs introduced in Chapter 4 and in this chapter, running them to process the FASTQ files we downloaded in the section "Copying the FASTQ Files."

First, we align the sequencing reads to the reference genome to produce a BAM file.

We will use the latest version of the human reference genome, called *GRCh38*. Some regions of some of the chromosomes have multiple possible sequences in GRCh38, referred to as *alternate contigs*. We won't use them, because it is not yet a settled matter how to apply annotations for interpretations to them. Thus, we will use the no-alt-contigs version of GRCh38.

We download the GRCh38 no-alt-contigs reference sequence to our EFS storage.

```
mkdir -p /home/ubuntu/efs/data/reference_data/reference_genome/
GRCh38_no_alt_contigs
cd /home/ubuntu/efs/data/reference_data/reference_genome/
GRCh38_no_alt_contigs
```

```
wget ftp://ftp.ncbi.nlm.nih.gov/genomes/all/GCA/000/001/405/
GCA_000001405.15_GRCh38/s
eqs_for_alignment_pipelines.ucsc_ids/GCA_000001405.15_GRCh38_no_alt_
analysis_set.fna.gz
```

You will see the download of this reference genome:

```
--2022-03-13 06:23:51--
ftp://ftp.ncbi.nlm.nih.gov/genomes/all/GCA/000/001/405/GCA_000001405.15_
GRCh38/seqs_for_alignment_pipelines.ucsc_ids/GCA_000001405.15_GRCh38_no_
alt_analysis_set.fna.gz
           => 'GCA_000001405.15_GRCh38_no_alt_analysis_set.fna.gz'
Resolving ftp.ncbi.nlm.nih.gov (ftp.ncbi.nlm.nih.gov)... 130.14.250.11,
130.14.250.10, 2607:f220:41e:250::12, ...
Connecting to ftp.ncbi.nlm.nih.gov (ftp.ncbi.nlm.nih.
gov)|130.14.250.11|:21... connected.
Logging in as anonymous ... Logged in!
==> SYST ... done.    ==> PWD ... done.
==> TYPE I ... done.  ==> CWD (1)
/genomes/all/GCA/000/001/405/GCA_000001405.15_GRCh38/seqs_for_alignment_
pipelines.ucsc_ids ... done.
==> SIZE GCA_000001405.15_GRCh38_no_alt_analysis_set.fna.gz ... 872949833
==> PASV ... done.    ==> RETR GCA_000001405.15_GRCh38_no_alt_analysis_
set.fna.gz ... done.
Length: 872949833 (833M) (unauthoritative)

GCA_000001405.15_GRCh38_no_alt_analysis 100%[=====================
========================================================>] 832.51M
58.7MB/s    in 17s

2022-03-13 06:24:08 (50.4 MB/s) - 'GCA_000001405.15_GRCh38_no_alt_
analysis_set.fna.gz' saved [872949833]
```

We need to create various indexes for this reference genome, because the programs we will run will need the indexes. We will use programs in the Docker container to do the indexing. So, let's go back into the container, making sure to bind our outside data directory to a directory in the container.

```
sudo chmod 666 /var/run/docker.sock
sudo service docker stop
sudo service docker start
docker run -v /home/ubuntu/efs/data:/home/data -it broadinstitute/
genomes-in-the-cloud:2.3.1-1512499786
```

Let's create the Samtools index.

```
cd /home/data/reference_data/reference_genome/GRCh38_no_alt_contigs
gunzip GCA_000001405.15_GRCh38_no_alt_analysis_set.fna.gz
samtools faidx GCA_000001405.15_GRCh38_no_alt_analysis_set.fna
```

When this has finished running, performing `ls -l` will show the newly created index file. It ends with `.fai`.

```
-rw-rw-r-- 1 1000 1000 3144230986 Mar 13 06:24
GCA_000001405.15_GRCh38_no_alt_analysis_set.fna
-rw-r--r-- 1 root root        7804 Mar 13 06:45
GCA_000001405.15_GRCh38_no_alt_analysis_set.fna.fai
```

Next we will create the BWA indexes. This operation will need more memory than what is available in our t2.micro EC2 instance, and so we will first upgrade the instance to t2.large, which has more memory. From the instance console, stop the t2.micro instance, choose Actions, then Instances Settings, and then select Change Instance Type (see Figure 8.41).

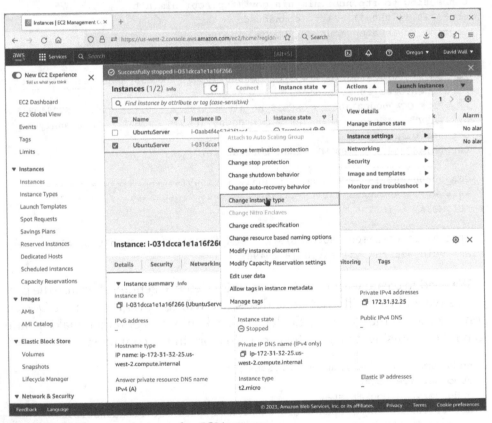

Figure 8.41: Changing the type of an EC2 instance

For Instance Type, choose t3.large, which has 8 GB of memory, and click the Apply button (see Figure 8.42).

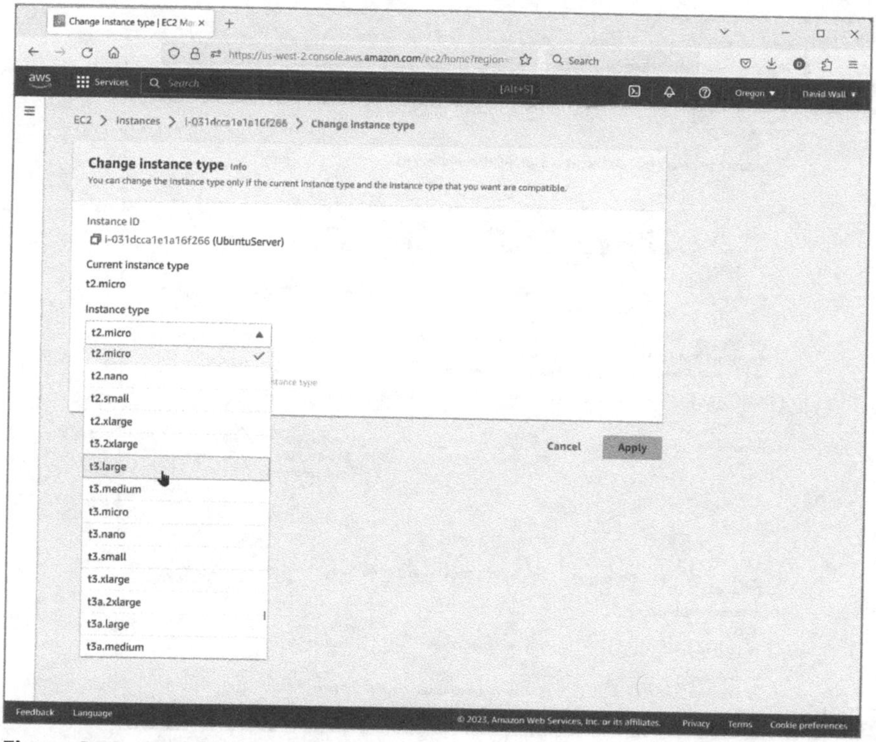

Figure 8.42: Applying the type change

Back in the EC2 instances console, click the browser refresh button (circular arrow on the left of the URL as the top of the screen) to see that the instance type of the EC2 instance is now t3.large (see Figure 8.43). Start the instance again.

Using the same key pair for the instance and using the new IP address shown in Public IPv4 Address, use puTTY to enter the command line of the instance again. Also, run the `efs mount`, `chmod`, and `service docker` commands that we need to run every time we stop and start the instance.

Create directories to store the BWA output files, which will be the output BAM file and the log file.

```
mkdir -p /home/ubuntu/efs/data/bam
mkdir -p /home/ubuntu/efs/data/logs
```

The next step is an indexing step and will take over an hour to run. You probably don't want to have your puTTY terminal open and connected to the EC2 instance over the Internet for a long time. We will use the `screen` facility to run long-running commands inside the container running inside a screen (inside our EC2 instance). This will allow us to detach from the screen and disconnect from the Internet and then later reattach the screen to see how the program in the container is running. Type **exit** to exit the container, and from the EC2 prompt type **screen**.

A welcome screen appears. Press Enter to remove it.

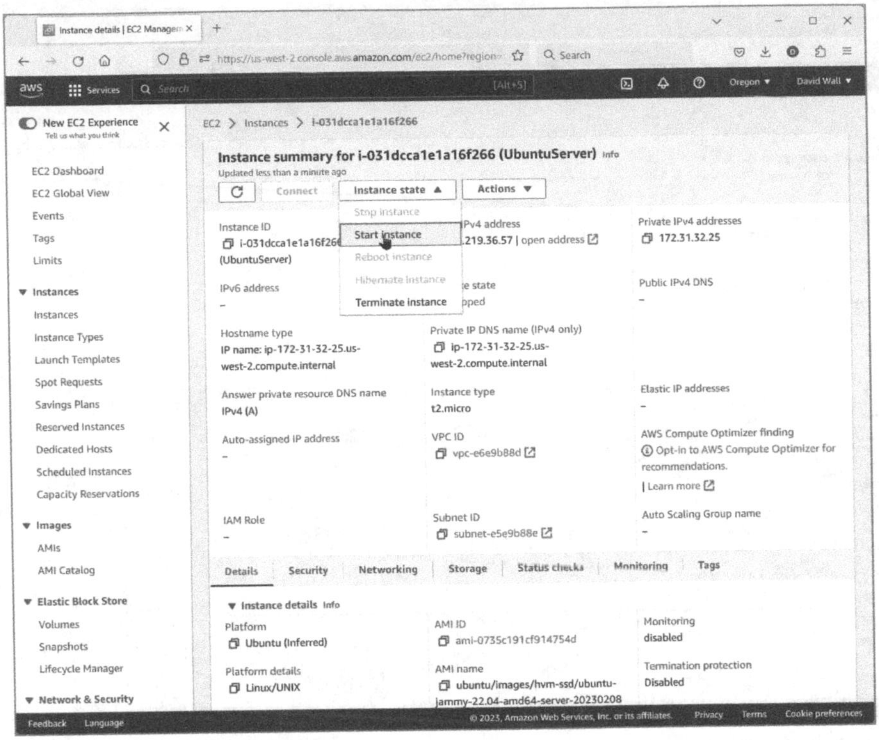

Figure 8.43: Restarting the modified EC2 instance

Re-enter the `broadinstitute/genomes-in-the-cloud` container, making sure to bind the `efs/data` with the `-v` option.

```
docker run -v /home/ubuntu/efs/data:/home/data -it broadinstitute/
genomes-in-the-cloud:2.3.1-1512499786
```

Inside the container, let's create the BWA indexes that will be used by the BWA program.

```
cd /home/data/reference_data/reference_genome/GRCh38_no_alt_contigs
/usr/gitc/bwa index GCA_000001405.15_GRCh38_no_alt_analysis_set.fna
```

You will start seeing output messages showing that the BWA index creation is happening.

```
[bwa_index] Pack FASTA... 25.78 sec
[bwa_index] Construct BWT for the packed sequence...
[BWTIncCreate] textLength=6199845082, availableWord=448243540
[BWTIncConstructFromPacked] 10 iterations done. 99999994 characters
processed.
[BWTIncConstructFromPacked] 20 iterations done. 199999994 characters
processed.
[BWTIncConstructFromPacked] 30 iterations done. 299999994 characters
processed.
```

Let's detach the screen so that we don't need to be connected via the Internet while this is running. To detach the screen, press Ctrl+A, then release them, and press the D key for detach. You will see a detach message such as this one:

```
[detached from 2187.pts-0.ip-172-31-93-231]
```

Take note of the screen number (which is 2187 in this example). We will need that later to reattach to the screen. If you forget the screen number, you can find it using the following command:

```
screen -ls
```

To reattach the screen to see how the processing is going, type the following, using the screen number (which in this example is 2187):

```
screen -r 2187
```

You will continue seeing output such as this:

```
[BWTIncConstructFromPacked] 660 iterations done. 6134031130 characters
processed.
[BWTIncConstructFromPacked] 670 iterations done. 6160559482 characters
processed.
[BWTIncConstructFromPacked] 680 iterations done. 6184133946 characters
processed.
[bwt_gen] Finished constructing BWT in 688 iterations.
[bwa_index] 3210.08 seconds elapse.
[bwa_index] Update BWT... 18.99 sec
[bwa_index] Pack forward-only FASTA... 16.86 sec
[bwa_index] Construct SA from BWT and Occ...
1493.23 sec
[main] Version: 0.7.15-r1140
[main] CMD: /usr/gitc/bwa index GCA_000001405.15_GRCh38_no_alt_
analysis_set.fna
[main] Real time: 4947.875 sec; CPU: 4767.848 sec
```

When the BWA indexing has finished running, typing `ls -lh` shows the new index files created by BWA,

```
total 8.8G
-rw-rw-r-- 1 1000 1000 3.0G Mar 13 06:24 GCA_000001405.15_GRCh38_no_alt_
analysis_set.fna
-rw-r--r-- 1 root root  18K Mar 13 13:28 GCA_000001405.15_GRCh38_no_alt_
analysis_set.fna.amb
-rw-r--r-- 1 root root  28K Mar 13 13:28 GCA_000001405.15_GRCh38_no_alt_
analysis_set.fna.ann
-rw-r--r-- 1 root root 2.9G Mar 13 13:28 GCA_000001405.15_GRCh38_no_alt_
analysis_set.fna.bwt
-rw-r--r-- 1 root root 7.7K Mar 13 06:45 GCA_000001405.15_GRCh38_no_alt_
analysis_set.fna.fai
-rw-r--r-- 1 root root 740M Mar 13 13:28 GCA_000001405.15_GRCh38_no_alt_
analysis_set.fna.pac
```

```
-rw-r--r-- 1 root root 740M Mar 13 10:46 GCA_000001405.15_GRCh38_no_alt_
analysis_set.fna.rpac
-rw-r--r-- 1 root root 1.5G Mar 13 13:54 GCA_000001405.15_GRCh38_no_alt_
analysis_set.fna.sa
```

We now have all the files to run BWA to align the FASTQ sequencing reads to the reference genome to produce a BAM alignment file.

Set the following variables that will be used for calling the BWA program. Our t3.large EC2 instance has two CPUs, so we tell BWA that there are two CPUs available by setting NCPUS to 2. If you are using your own FASTQ sequencing paired read files instead of the ones we downloaded, then enter them for infile1 and infile2. If you have more than one pair of FASTQ files, then you will run BWA for each pair, and each BWA run will have different values for ID (flowcell and lane), PU (flowcell, lane, and sample), and/or LB (library prep identifier, which could be a date). This will allow the future quality score recalibration step to do a better job. If you have only one pair of FASTQ files that contains all the sequencing read data for the sample, then the values in the ID, PU, and LB fields are not important given that there is no need to distinguish read from the different FASTQ files from different sequencing runs. The SM field contains the sample identifier that will be carried through to the end of variant processing, so choose a value that identifies the sample.

```
cd /home/data/bam
outfile=/home/data/bam/my_bam.bam
infile1=/home/data/fastq/SRR622457_1.fastq.gz
infile2=/home/data/fastq/SRR622457_2.fastq.gz
ref_fasta=/home/data/reference_data/reference_genome/GRCh38_no_alt_
contigs/GCA_000001405.15_GRCh38_no_alt_analysis_set.fna
NCPUS=2
ID=SRR622457:1
PU=SRR622457:1:NA12878
LB=20200101
PL=ILLUMINA
SM=NA12878
read_group_id="@RG\tID:$ID\tPU:$PU\tLB:$LB\tPL:$PL\tSM:$SM"
log_file=/home/data/logs/bwa_output.txt
```

Let's start running the bwa command to align our sequencing reads to the reference genome to produce the BAM alignment file.

```
/usr/gitc/bwa mem -t $NCPUS -R "$read_group_id" "$ref_fasta" "$infile1"
"$infile2" | samtools view -1 - -o $outfile
You will see that bwa mem has started processing the FASTQ data.
[M::mem_pestat] skip orientation FF as there are not enough pairs
[M::mem_pestat] analyzing insert size distribution for orientation FR...
[M::mem_pestat] (25, 50, 75) percentile: (368, 390, 410)
[M::mem_pestat] low and high boundaries for computing mean and std.dev:
(284, 494)
[M::mem_pestat] mean and std.dev: (391.53, 36.22)
[M::mem_pestat] low and high boundaries for proper pairs: (242, 536)
[M::mem_pestat] skip orientation RF as there are not enough pairs
```

```
[M::mem_pestat] skip orientation RR as there are not enough pairs
[M::mem_process_seqs] Processed 198020 reads in 93.953 CPU sec, 47.687
real sec
[M::process] read 198020 sequences (20000020 bp)...
[M::mem_pestat] # candidate unique pairs for (FF, FR, RF, RR): (0,
102, 0, 0)
[M::mem_pestat] skip orientation FF as there are not enough pairs
[M::mem_pestat] analyzing insert size distribution for orientation FR...
[M::mem_pestat] (25, 50, 75) percentile: (356, 387, 413)
[M::mem_pestat] low and high boundaries for computing mean and std.dev:
(242, 527)
```

This BWA mem program will take hours to run, so you may want to detach the screen by pressing Ctrl+A at the same time and then pressing D. You may want to quit the puTTY terminal and come back to it later. When you open the puTTY terminal in a few hours' time, you can see that BWA has started creating your BAM file from the FASTQ files. Type the following:

```
ls -lh /home/ubuntu/efs/data/bam
```

and you will see something similar to the following, showing that the BAM file is being created and was last written to a few minutes ago:

```
-rw-r--r-- 1 root root 3.2G Mar 13 20:10 my_bam.bam
```

If you use the top command to inspect CPU usage, you will see something like the following, showing that program BWA is using 200 percent of the CPUs. Our t3.large EC2 instance has two CPUs. The 200 percent tells us that BWA is using 100 percent of each of the two CPUs.

```
Tasks: 143 total,   3 running, 140 sleeping,   0 stopped,   0 zombie
%Cpu(s): 30.3 us,  0.2 sy,  0.0 ni,  0.0 id,  0.0 wa,  0.0 hi,  0.0 si, 69.6 st
MiB Mem :   7849.9 total,    118.5 free,   6211.4 used,   1520.0 buff/cache
MiB Swap:      0.0 total,      0.0 free,      0.0 used.   1366.4 avail Mem

    PID USER      PR  NI    VIRT    RES    SHR S  %CPU  %MEM     TIME+ COMMAND
   4364 root      20   0 6329716   5.8g   2720 S 200.0  75.5 778:47.39 bwa
   6864 ubuntu    20   0   11016   3776   3228 R   2.3   0.0   0:00.19 top
     12 root      20   0       0      0      0 S   0.3   0.0   0:01.11 ksoftirqd/0
      1 root      20   0  101916  10484   7352 S   0.0   0.1   0:06.70 systemd
      2 root      20   0       0      0      0 S   0.0   0.0   0:00.01 kthreadd
      3 root       0 -20       0      0      0 I   0.0   0.0   0:00.00 rcu_gp
      4 root       0 -20       0      0      0 I   0.0   0.0   0:00.00 rcu_par_gp
      6 root       0 -20       0      0      0 I   0.0   0.0   0:00.00 kworker/
0:0H-events_highpri
      9 root       0 -20       0      0      0 I   0.0   0.0   0:00.00 mm_percpu_wq
     10 root      20   0       0      0      0 S   0.0   0.0   0:00.00 rcu_
tasks_rude_
     11 root      20   0       0      0      0 S   0.0   0.0   0:00.00 rcu_
tasks_trace
     13 root      20   0       0      0      0 R   0.0   0.0   0:01.16 rcu_sched
     14 root      rt   0       0      0      0 S   0.0   0.0   0:00.17 migration/0
```

15 root	-51	0	0	0	0 S	0.0	0.0	0:00.00	idle_
inject/0									
16 root	20	0	0	0	0 S	0.0	0.0	0:00.00	cpuhp/0
17 root	20	0	0	0	0 S	0.0	0.0	0:00.00	cpuhp/1
18 root	-51	0	0	0	0 S	0.0	0.0	0:00.00	idle_
inject/1									
19 root	rt	0	0	0	0 S	0.0	0.0	0:00.21	migration/1
20 root	20	0	0	0	0 R	0.0	0.0	0:00.61	ksoftirqd/1
22 root	0	-20	0	0	0 I	0.0	0.0	0:00.00	kworker/
1:0H-events_highpri									
23 root	20	0	0	0	0 S	0.0	0.0	0:00.00	kdevtmpfs
24 root	0	-20	0	0	0 I	0.0	0.0	0:00.00	netns
25 root	0	-20	0	0	0 I	0.0	0.0	0:00.00	inet_frag_wq
26 root	20	0	0	0	0 S	0.0	0.0	0:00.00	kauditd
28 root	20	0	0	0	0 S	0.0	0.0	0:00.01	khungtaskd
29 root	20	0	0	0	0 S	0.0	0.0	0:00.00	oom_reaper
30 root	0	-20	0	0	0 I	0.0	0.0	0:00.00	writeback
31 root	20	0	0	0	0 S	0.0	0.0	0:01.48	kcompactd0
32 root	25	5	0	0	0 S	0.0	0.0	0:00.00	ksmd
33 root	39	19	0	0	0 S	0.0	0.0	0:00.00	khugepaqed
79 root	0	-20	0	0	0 I	0.0	0.0	0:00.00	kintegrityd
80 root	0	-20	0	0	0 I	0.0	0.0	0:00.00	kblockd
81 root	0	-20	0	0	0 I	0.0	0.0	0:00.00	blkcg_
punt_bio									
82 root	0	-20	0	0	0 I	0.0	0.0	0:00.00	tpm_dev_wq
83 root	0	-20	0	0	0 I	0.0	0.0	0:00.00	ata_sff
84 root	0	-20	0	0	0 I	0.0	0.0	0:00.00	md
85 root	0	-20	0	0	0 I	0.0	0.0	0:00.00	edac-poller

Press Ctrl+C to quit the `top` command.

When the BWA program finishes, the file `my_bam.bam` will contain all the alignments of the reads to the reference genome. The next steps (refinement of the BAM file, calling, filtering, annotating, and interpreting the variants) are conducted in the same way.

Summary

This chapter discussed the process by which we get from raw sequencing data to useful information about a genome. We built a pipeline to align a sample to a genome reference, refine the alignment data, call variants (SNVs, indels, SVs, CNVs), annotate and prioritize the variants, and perform inheritance analysis among family members. This was done both generically—in an arbitrary Linux environment—and in the AWS cloud specifically.

The next chapter will show how to visualize genomic data using a genome browser. We will cover how to visualize the alignments to the reference genome, as well as how to integrate public data on additional tracks to get a better understanding of the genome.

Visualizing the Genome

In the previous chapter, we performed quality control steps on the FASTQ file delivered by the sequencing facility, and we aligned all the sequences of that file to the human genome reference. We are only halfway through our analysis, as we still have to, first, "call the variants" (that is, formally identify all the differences between our genome and the genome reference) and, second, annotate these variants in order to find out their impact—such as, do they change the protein encoded by a gene? Are they common or rare? Benign or potentially pathogenic? And so forth.

Before moving to these next steps, it is helpful to visualize the result of the alignment of our genome to the human genome reference. It will help us to better understand the process of genome analysis, as well as how our genome works. We will use a genome visualizer for this step.

Introducing Genome Visualizers

What is a genome visualizer? It is a graphical tool that allows you to visualize genomic data. The sequence of a genome is usually displayed as a horizontal line. Genomic data associated with that sequence is aligned to the sequence and organized as "tracks" above or below the genome sequence. For example, tracks can include RNA or protein information, DNA modifications, transcription factors binding sites, degree of complexity of the sequence, degree of conservation of the sequence versus other species, or alignments of sequenced reads to that genome sequence.

Some genome visualizers are particularly suited to browsing publicly available data, although they also allow you to import and visualize your own private genomic data. These genome visualizers include the UCSC Genome Browser (genome.ucsc.edu), the Ensembl Genome Browser (www.ensembl.org), or the NCBI Genome Workbench (www.ncbi.nlm.nih.gov/tools/gbench). The UCSC Genome Browser is particularly popular, with several million hits a day. Many large international projects, like the ENCODE, FANTOM, or GTEx projects, make their data public as tracks in this browser. The ENCODE and FANTOM projects seek to describe the regulatory elements and transcriptome of the human genome; the GTEx project analyzed the transcriptome and expression quantitative trait loci (eQTL) of 54 types of tissues from 1,000 donors. In addition, the journal *Nucleic Acids Research* requires authors to create a track for reviewers, and some authors make these tracks available to the public. The UCSC Genome Browser is also available as a desktop application, called Genome Browser in a Box (GBiB).

Figure 9.1 displays a view of the UCSC Genome Browser for the tumor necrosis factor (TNF) gene, which when expressed promotes systemic inflammation. The TNF protein is targeted by several inhibitors to treat rheumatoid arthritis, including adalimumab, which is currently the largest selling pharmaceutical drug in the world.

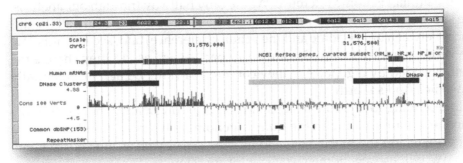

Figure 9.1: The UCSC Genome Browser displaying the tumor necrosis factor gene

From top to bottom, this TNF view shows the following:

- The location of the TNF gene on chromosome 6 (small vertical bar).
- The DNA structure of the TNF gene: we can see the 5′UTR, first exon (striped box), long first intron (dotted line), small second exon (striped box), and beginning of second intron of the TNF gene.
- The corresponding mRNA. As expected, the introns have been spliced out.

Other tracks have been added here:

- Hypersensitive DNase regions. These correspond to areas where the chromatin is open and that therefore are ready to be bound by regulatory proteins such as transcription factors.
- Conserved regions across species. We can visualize that exons are more conserved than introns.
- Common variants from the database dbSNP.
- RepeatMasker track, which are areas with repeats or low-complexity sequences.

From the scale indicated just below the chromosome view in Figure 9.1 (which shows the UCSC Genome Browser view of the TNF gene), we can see that this view displays about 1000 bp. If we zoom in, we can visualize the sequence of the gene. Figure 9.2 displays the third exon of the TNF gene, with the DNA sequence and the resulting protein sequence.

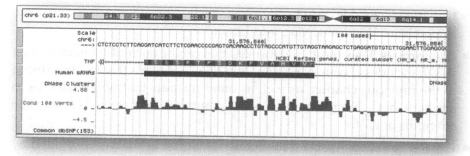

Figure 9.2: The third exon of the TNF gene

We have seen so far that genome browsers allow us not only to visualize genomes but also to provide important contextual information to understand how the genome works. This information will be crucial to help infer what could be the consequences of variations in our genome.

For visualizing and interpreting the alignment of your private sequenced reads to a reference genome, most prefer to use desktop genome visualizers such as Integrative Genomics Viewer (IGV; www.broadinstitute.org/igv), Integrated Genome Browser (IGB; bioviz.org), JBrowse (jbrowse.org), Tablet (ics.hutton.ac.uk/tablet), or Artemis (www.sanger.ac.uk/tool/artemis). IGB was released in 2001 and is one of the oldest visualizers; its architecture is showing its age and its development has stalled, so we would not recommend starting with this browser. JBrowse is a powerful and very customizable browser with many plugins and can run as a desktop application or be web-based; its configuration can be quite complex, and it is best suited for developers wanting to embed genomic views in their websites. Tablet is a lightweight visualizer that is easy to use but is restricted in the supported file formats and connections to external resources. Artemis is convenient for small genomes such as bacterial genomes but can struggle with larger genomes like human or other eukaryotic genomes or with a large number of tracks. IGV is a very comprehensive and customizable genome browser. It can handle a large variety of file formats and offers a lot of functionalities, while remaining easy to use. For these reasons, IGV is a popular genome visualizer and will be our genome visualizer of choice in this chapter. A web version of IGV, called IGV Web App, was released in 2018. IGV Web App can be useful for sharing genome views with other people. In this chapter, we will focus on the desktop version of IGV.

Installing the IGV Desktop Visualizer

Go to the IGV desktop main page at `www.broadinstitute.org/igv`. The menu on the left contains three useful links, as shown in Figure 9.3.

- **Downloads:** Enables you to download the latest version of the visualizer, available for Windows, Mac, and Linux.

- **File formats:** Lists the 40 file formats that can be currently visualized on IGV. It includes of course the BAM and FASTA formats that we introduced in the previous chapter, as well as other formats that will enable us to annotate our genomic data. This large number of formats is one of the strengths of IGV.

- **Hosted genomes:** Lists all the reference genomes that are provided by default by IGV. This list covers several builds of the human genome, as well as the genomes of other species such as a selection of mammals, fish, insects, fungi, plants, worms, protozoans, bacteria, and viruses.

Figure 9.3: The IGV desktop main page

Click Downloads to access the IGV installation links for Windows, Mac, and Linux. Where should we install IGV? We could install IGV in our AWS EC2 instance and visualize genomes from our Windows desktop by using a graphical terminal emulator like PuTTY with an X server for Windows, as discussed in Chapter 7. This would mean all the graphical IGV data would transit over the network, which would be slow and with a lower resolution and defeat the purpose of a local genome browser. We will instead install IGV on our desktop (or laptop) and enable a back-end connection between our desktop and our genomic data, which resides on AWS.

We like to have consistent environments between our desktop and our Linux EC2 instance and use a Linux virtual machine (for example, an Ubuntu virtual machine with VMware, as in Chapter 7) to access AWS. So we will install the Linux version of IGV in our Linux virtual machine:

1. Open a web browser window in your Linux virtual machine.

2. Go to www.broadinstitute.org/igv.

3. Click Downloads and then the IGV For Linux icon.

4. Click Save File.

5. Unzip the downloaded file to our installation directory (/opt).

```
Sudo unzip /home/cv/Downloads/IGV_Linux_2.8.7.zip  -d /opt
```

The downloaded files automatically include the prerequisite Java 11, so no additional software needs to be installed. To launch the IGV browser, launch the already provided shell script igv.sh.

This script starts IGV with 4 GB of memory, which is sufficient for most uses. For large files with many tracks, the memory can be increased to, for example, 8 GB, by editing the igv.sh script. Replace this:

```
"java --module-path=lib -Xmx4g"
```

with this:

```
"java --module-path=lib -Xmx8g"
```

Let's launch the IGV browser.

```
cd /opt/IGV_Linux_2.8.7
./igv.sh
```

The window of the IGV browser opens, as shown in Figure 9.4.

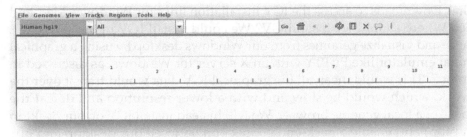

Figure 9.4: The appearance of the IGV browser on initial startup

We now need to do the following:

1. Select a genome reference. By default, the GRCh37/hg19 reference is selected, but we can select other references on the server provided by IGV or load our own genome reference.

2. Load our aligned genome, which is the BAM file we generated in Chapter 8.

Connecting the IGV Visualizer to Our AWS Data

The IGV Browser can load data from the following:

- Local files
- URLs, which allows files to be loaded from HTTP or FTP servers
- AWS or other cloud providers, as well as remote servers

Importantly, the IGV browser can visualize remote files without downloading the whole file.

How can we connect our IGV desktop browser to our BAM file located on our AWS S3 bucket?

One method consists of making our BAM file public on AWS, generating a public AWS URL, and typing in this URL in the IGV browser.

1. Click to select your BAM file in your S3 bucket.

2. Click the Actions menu, and select Make Public.

3. Get the URL of the file in S3: click to select the file, click the Properties button, and copy the contents of the Link field. Save the URL for use with the IGV browser.

It is often preferable, however, to keep our data private. So an easy alternative is to mount, that is connect, our S3 bucket to our desktop so that our S3 bucket will be accessible like a local drive. We will use for this the s3fs software, which runs on Linux and macOS.

If you are working on Ubuntu 16.04 or newer, use this:

```
sudo apt install s3fs
```

We can check that s3fs installed correctly as follows:

```
which s3fs
-> /usr/bin/s3fsst
s3fs
-> s3fs: missing BUCKET argument.
   Usage: s3fs BUCKET:[PATH] MOUNTPOINT [OPTION]...
```

This is the expected message when typing the s3fs command without specifying the arguments, so s3fs installed correctly.

On macOS, s3fs is best installed via Homebrew.

```
brew cask install osxfuse
brew install s3fs
```

We now need to obtain security credentials—an access key ID and a secret access key—from AWS.

1. Go to the AWS Management Console (aws.amazon.com/console) and sign in as a root user.

2. In the AWS menu bar, click your user ID and then select My Security Credentials in the drop-down menu. Click Continue To Security Credentials.

3. Click Users in the left IAM menu, and then click Add User.

4. Type in a username of your choice, and select Programmatic Access as the access type (see Figure 9.5).

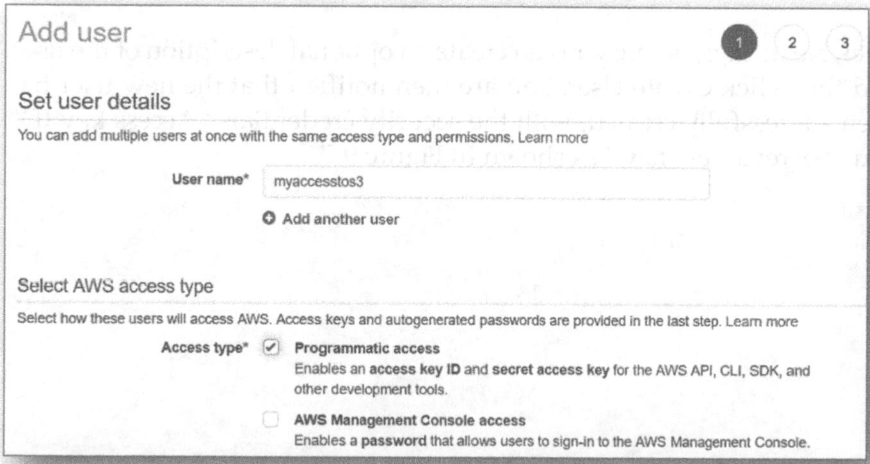

Figure 9.5: Adding a user in an AWS account

5. The next panel prompts you to set the user permissions. You can click Attach Existing Policies Directly and select the policy AmazonS3FullAccess (as shown in Figure 9.6), or you can create a new group with this policy and add the new user to this group. The policy AmazonS3FullAccess will give the user the permission to execute PUT, GET, and DELETE objects on the S3 bucket. You can also use the policy AmazonS3ReadOnlyAccess or create a custom policy if you want to be more restrictive.

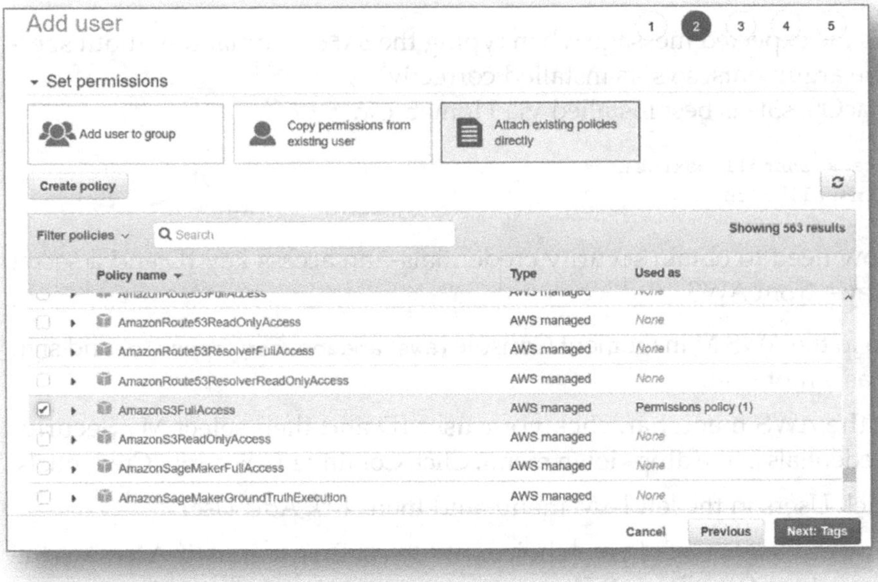

Figure 9.6: Granting S3 storage permissions to the newly created user

6. Click Next: Tags, where you can create an optional description of the user, and then click Create User. You are then notified that the new user has been successfully created, with the security credentials "Access key ID" and "Secret access key," as shown in Figure 9.7.

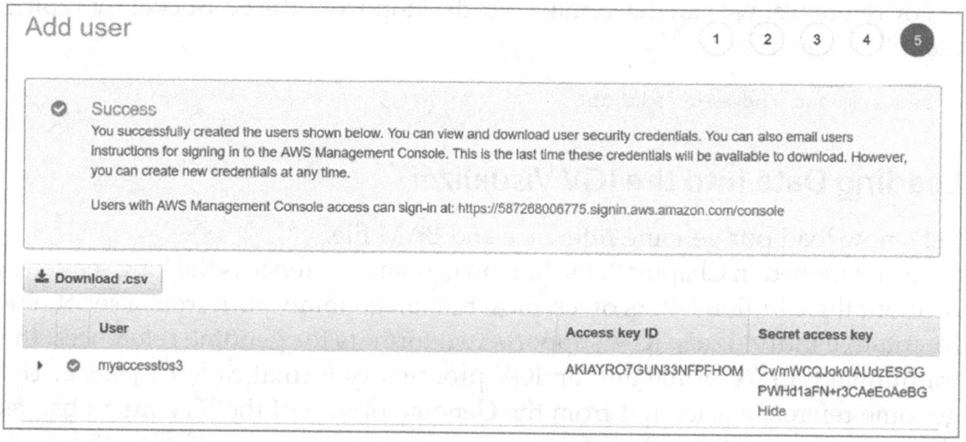

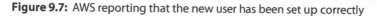

Figure 9.7: AWS reporting that the new user has been set up correctly

7. Copy your access key and secret access key, and save them in a file on your desktop using the following command:

```
#echo access_key_ID:secret_access_key  > /home/cv/.credentials-s3fs
echo AKIAYRO7GUN33NFPFHOM:Cv/mWCQJok01AUdzESGGPWHd1aFN+r3CAeEoAeBG >
/home/cv/.credentials-s3fs
```

8. Change the permission of the new file to remove access for others.

```
chmod 600 /home/cv/.credentials-s3fs
```

9. Create a directory for connecting to the S3 bucket.

```
mkdir /home/cv/s3mount
```

10. Connect the Ubuntu desktop to the S3 bucket mys3bucket.

```
s3fs mys3bucket /home/cv/s3mount -o passwd_file=/home/cv/.
credentials-s3fs
```

To check that the desktop is properly connected to the S3 bucket, type the following:

```
df -h
-> Filesystem     Size  Used Avail Use%   Mounted on
   s3fs           256T  0    256T  0      /home/cv/s3mount
```

You can now access all the files of your S3 bucket as if they were local files in your /home/cv/s3mount directory. You can, for example, list all the files of your S3 bucket with this:

```
ls /home/cv/s3mount/*.*
```

For reference, we can disconnect our desktop from the S3 bucket by typing the following:

```
sudo umount /home/cv/s3mount
```

Loading Data into the IGV Visualizer

Let's now load our genome reference and BAM file.

As mentioned in Chapter 8, the human genome reference is the long sequence of about three billion letters of a typical human genome, built from a set of representative individuals. IGV supports two formats for genome references: the traditional FASTA format and an IGV proprietary format called .genome. The genome reference is loaded from the Genome option of the IGV menu bar, as shown in Figure 9.8.

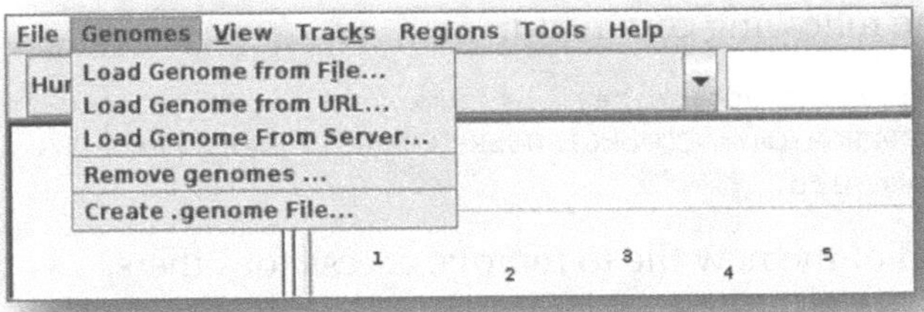

Figure 9.8: Opening a data file in the IGV browser

Since our S3 bucket is connected to our desktop and appears as the local directory s3mount, we can load our genome reference using Load Genome from File. We then double-click the s3mount directory and select our grch38.fa genome reference. IGV requires the genome reference to be indexed, so make sure the index file grch38.fai is copied in the same directory.

Alternatively, you can load the genome reference from the list of available genome references hosted on the IGV server. Click Load Genome From Server to list all the genomes hosted by the IGV server. Select Human hg38 and click OK. This will add this genome reference to the top-left drop-down list of available genome references. Selecting Download Sequence is optional, as the human genome reference is about 3 GB; IGV can work with a remote genome reference and fetch parts of the remote sequence as necessary, as shown in Figure 9.9.

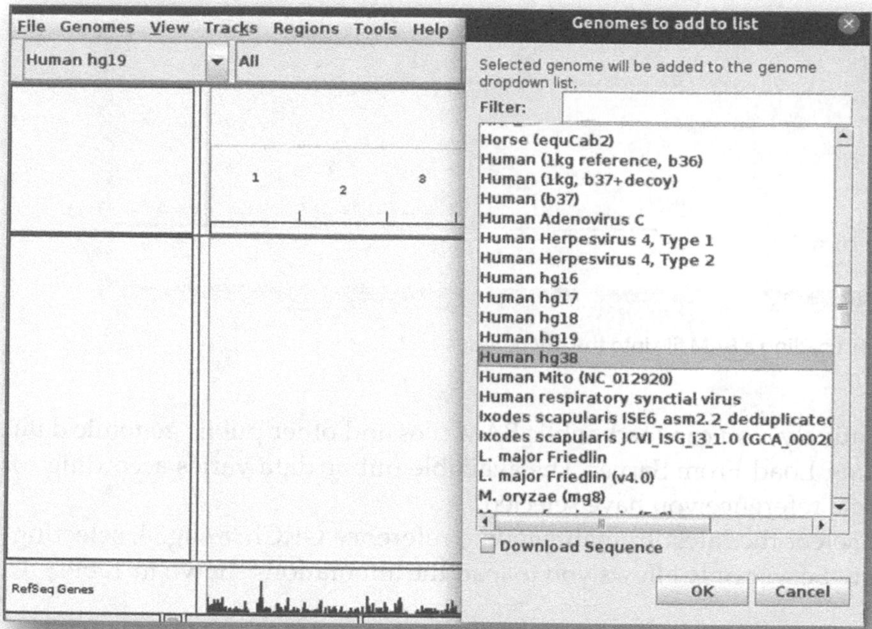

Figure 9.9: Choosing a genome reference

The GRCH38/hg38 can now be selected from the genome reference drop-down list. Click it to load the sequence, as shown in Figure 9.10.

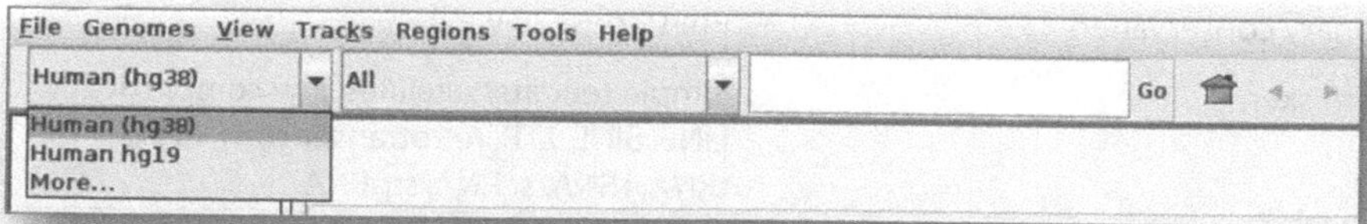

Figure 9.10: Specifying the hg38 human genome as a reference

We can now load our BAM file by clicking the File option of the IGV menu bar, selecting Load From File, double-clicking the s3mount directory, and selecting the file seqpaired.sorted.bam we generated in the previous chapter. IGV requires the BAM file to be sorted and indexed and the index (suffix BAM.BAI) to be present in the same directory, as shown in Figure 9.11.

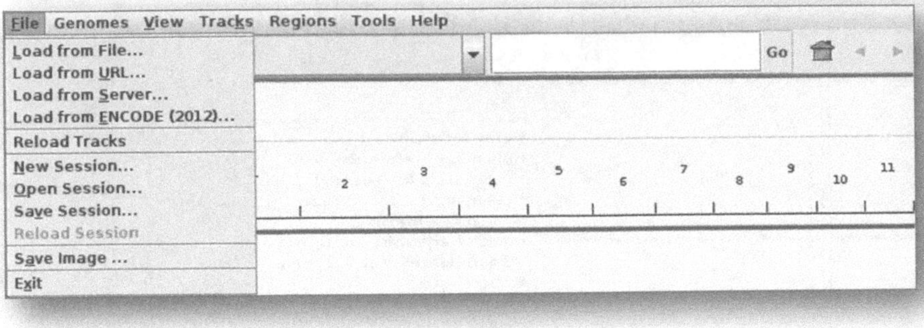

Figure 9.11: Loading a BAM file into the IGV browser

Alternatively, you can load public BAM files and other public genomic data by selecting Load From Server. The available public data varies according to the genome reference you have selected.

If you select the latest human genome reference GRCh38/hg38, selecting Load From Server only allows you to load the annotations shown in Table 9.1.

Table 9.1: Setting hg38 Options in the IGV Browser

GRCH38/HG38	
Category	**Options**
Gene annotations	Gencode V24, Gencode V24 Basic, Ensembl Genes (94)
Single Nucleotide Variants (SNVs)	All Snps 1.4.2, Common Snps 1.4.2
Evolutionary conservation	Phastcons (20 way)
Repeat masker	Simple repeats, satellites, low complexity, LINE, SINE, LTR, retrotransposons, RNA, tRNA, rRNA, scRNA, srpRNA

If you select the older GRCh37/hg19 human genome reference, selecting Load From Server allows you to load more extensive annotations, as well BAM files from the Platinum project (six family members sequenced at a depth of 50X), the 1000 Genomes projects (Exomes and Low coverage WGS), some files from The Cancer Genome Atlas (TCGA), and files from the ENCODE project (see Table 9.2).

Table 9.2: Setting hg19 Options in the IGV Browser

GRCH37/HG19	
Category	**Options**
Gene annotations	Ensembl Genes, UCSC Genes, MGC Genes, Gencode Genes (V10 to V18)
Phenotype & disease	GWAS Catalog, RGD Human QTL
Sequence & regulation	CpG islands, GC %
Variations & repeats	dbSNP 1.4.7, Genomic Structural Variation, Repeat Masker
Comparative Genomics	siphy rate (10 mer), siphy pi, Phastcons (Vertebrate 46 way), Multiple Alignments (44 species)
TCGA 2016	DNA Methylation, CopyNumber and Transcriptome (RNA-seq) for several cancer types.
ENCODE	Transcription factors binding (ChIP), DNase clusters, Transcriptome (RNA-seq) for 7 cell lines, CTCF binding, Histone marks for 9 cell lines
	The option "Load from Encode (2012)" offers access to additional files & formats.
1000 Genomes	Sample information, sites, genotypes, BAM files (exomes and low coverage WGS)
Platinum genomes	BAM files of 6 family members sequenced at a depth of 50X: NA12877, NA12878, NA12889, NA12890, NA12891, NA12892

All this data can be loaded as additional tracks below our aligned genome, and it allows us to interpret the impact of variants in our genome.

Given the large number of genomic files available for the older hg19 genome reference, it is a good idea to get proficient with IGV using the older hg19 reference and its associated files. It is important to note, however, that some of these files are out of date. Once you feel comfortable and ready to do serious work on your own genomic data, you can download the latest tracks data from the UCSC Table Browser or the UCSC FTP server. To access these, go to the UCSC home page (genome.ucsc.edu) and click either Table Browser under the Tools heading or FTP under the Downloads heading.

As shown in Figure 9.12, we use the UCSC Table Browser to download a recent track called RefSeq Func Elems for the GRCh38/hg38 build. This track highlights functional elements that are not genes and that have been experimentally validated. This functional element includes gene regulatory regions (such as enhancers or silencers of transcription), chromatin structural elements (like insulators, which induce the formation of DNA loops and rearrange the nucleosome, and DNase hypersensitive sites), DNA replication origins, and clinically significant areas where DNA tends to recombine or is instable.

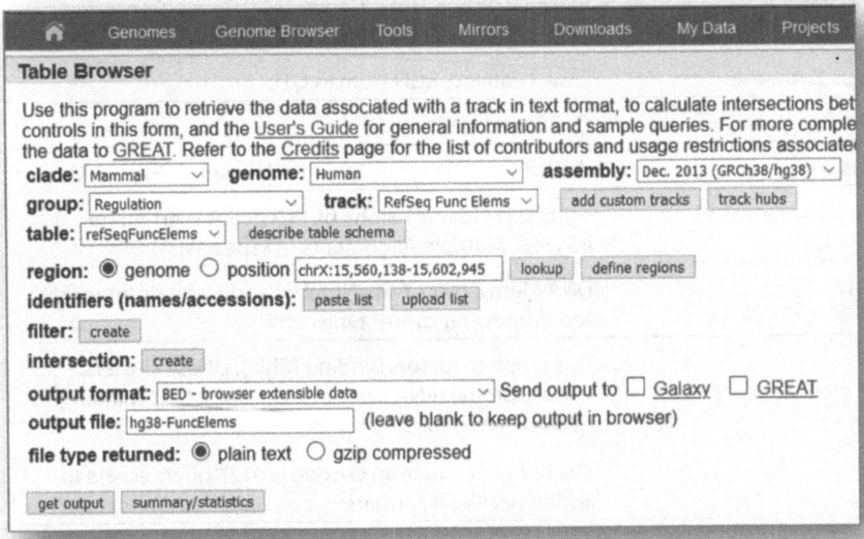

Figure 9.12: Setting up the UCSC Table Browser to download functional elements

To download the track, select the assembly GRCh38/hg38, the group of tracks called Regulation, and then the track REfSeq Func Elems. As an output format, the BED format, which stands for Browser Extensible Data, is a convenient format for exporting track information from UCSC and importing it into IGV. The BED format is extensively used in genome analysis to describe genomic areas. The BED format includes three mandatory fields: chrom (chromosome name), chromStart, and chromEnd (starting and ending position of the feature on the chromosome), as well as additional optional fields to help draw the line thickness, shade, and color of the feature if it is displayed as a track in a genome browser.

To identify which UCSC tracks can be particularly useful, it is a good idea to visualize these tracks in the UCSC Genome Browser. Go to the UCSC home page (genome.ucsc.edu) and click Genome Browser. By default, the human reference genome GRCH38/hg38 is selected. Click the GO button. This launches the UCSC Genome Browser, which currently displays by default a number of tracks for the ACE2 gene, which is the target and main port of entry of the SARS-CoV-2 virus (Figure 9.13). Figures 9.1 and 9.2 at the beginning of this chapter showed similar tracks for the TNF gene. Scroll down, and you can see a large number of additional track options. Click the hyperlink of one of these tracks to get a full description. In that full description page, change the display status of this track from hide to full, or any intermediary state between (dense/squish/pack), and hit Submit to display the track in the browser. Alternatively, change the display status directly on the browser page and hit Refresh to achieve the same result.

You can hover your mouse over the gray vertical bar at the left of each track to get the name of the track and click that bar to get a full explanation of the track (the black arrow in Figure 9.13). You can right-click anywhere on a track to change its display status.

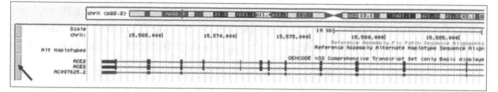

Figure 9.13: The UCSC Genome Browser showing the ACE2 gene

The number of tracks and parameters, as shown in Figure 9.14, can be overwhelming, and it can sometimes be hard to return to the default settings. Closing and restarting the UCSC Genome Browser does not help, as the settings of the last sessions are saved. The option Reset All User Settings in the drop-down menu of the main menu bar, shown in Figure 9.15, will save you a headache.

Phenotype and Literature					refresh
OMIM Alleles / dense	Cancer Gene Expr... / hide	ClinGen CNVs / hide	ClinVar Variants / hide	Coriell CNVs / hide	COSMIC Regions / hide
Development Delay / hide	Gene Interactions / hide	GeneReviews / hide	GWAS Catalog / hide	HGMD Variants / hide	LOVD Variants / hide
OMIM Cyto Loci / hide	OMIM Genes / hide	SNPedia / hide	TCGA Pan-Cancer / hide	UniProt Variants / hide	Variants in Papers... / hide

mRNA and EST					refresh
P12 Human mRNAs / hide	P12 Spliced ESTs / hide	P12 Human ESTs / hide	P12 Other ESTs / hide	P12 Other mRNAs / hide	SIB Alt-Splicing / hide

Expression					refresh
P12 GTEx Gene / pack	GTEx Transcript / hide	New GTEx Gene V8 / hide	Affy GNF1H / hide	Affy U133 / hide	Affy U95 / hide
EPDnew Promoters / hide	GNF Atlas 2 / hide	GWIPS-viz Riboseq / hide	miRNA Tissue Atlas / hide		

Regulation					refresh
P12 ENCODE Regulation... / show	GeneHancer / hide	P12 CpG Islands... / hide	New ENCODE cCREs / hide	Hi-C and Micro-C / hide	ORegAnno / hide
RefSeq Func Elems / hide					

Comparative Genomics					refresh
Conservation / full	Cons 7 Verts / hide	Cons 20 Mammals / hide	Cons 30 Primates / hide	Primate Chain/Net / hide	Placental Chain/Net / hide
Vertebrate Chain/Net / hide					

Variation					refresh
dbSNP 153 / pack	Common SNPs(151) / hide	Common SNPs(150) / hide	Common SNPs(147) / hide	Common SNPs(146) / hide	Common SNPs(144) / hide

Figure 9.14: User settings and preferences in the UCSC Genome Browser

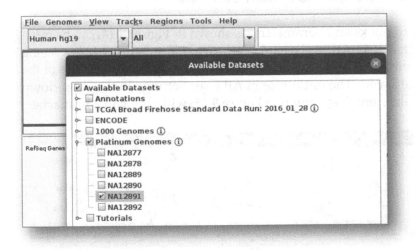

Figure 9.15: Resetting everything to default values

Visualizing Aligned Sequencing Reads in IGV

As a practice exercise, let's load the genome reference GRCh37/hg19 and the public BAM file NA12891, which corresponds to an American male, of European ancestry and resident in Utah, sequenced at a depth of 50x. This is shown in Figure 9.16.

Figure 9.16: Choosing a publicly available BAM file representing a population of men of European ancestry in Utah

After the data is loaded, the IGV window stays empty. It is normal, in order to not use too much memory, that the reads become visible only when we zoom in to reach a visualization span of 30 kbp. This threshold can be increased if you are working with low-coverage files, or decreased for files with a high read depth. To change the threshold, click View in the top menu bar, select Preferences in the drop-down menu, click the Alignment tab, and change Visibility Range Threshold from 30 kbp to your desired value. Beware that if the threshold is too high, your file might fail to load.

You can navigate the genome and zoom in and out by doing the following:

- Selecting a particular chromosome in the chromosome drop-down list (default: All)

- Typing in the search box: a gene name (for example: TNF), a locus with genomic coordinates (like chrX:15,597,575-15,599,197), or a mutation (for example, KRAS:G12C to specify an amino acid change, or KRAS:123A>T to indicate a nucleotide change)

- Pressing the + or - zoom controls at the top-right corner of the IGV window

- Double-clicking anywhere in a track to zoom in, and double-clicking while pressing the Alt key to zoom out

- Clicking and dragging the mouse on the genome ruler to select an area to which to zoom

- Clicking on the home button (just right of the Go button) to return to the whole view of the genome

Note that you can click anywhere on a chromosome ideogram to jump to that location.

Figure 9.17 shows the IGV window after typing ACE2 in the search box and double-clicking a few times in the reads track to reach 30kbp in the genome ruler.

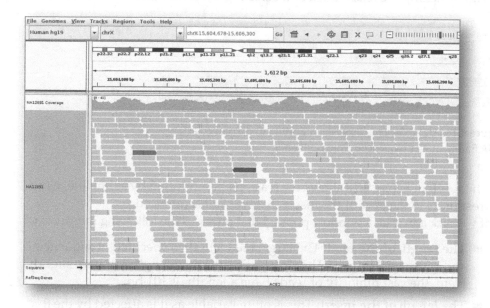

Figure 9.17: Navigating a genome in the UCSC Genome Browser

From top to bottom in Figure 9.17, the features of the IGV window are as follows:

- An ideogram of the chromosome X, and a vertical red bar noting our location on this chromosome.

- A genome ruler that indicates the number of bp spanned along the width of the window (1612 bp here), and the genomic coordinates.

- A coverage track, which represents the number of aligned reads at each genomic location. You can click this track at several genomic coordinates. Each time, a window will pop up and indicate the genomic coordinate, the total number of aligned reads at that position, and the number and proportion of A, T, G, and C.

- A reads track that shows how the reads align to the genome reference.

We can see that all the reads have an orientation, pointing to the left or to the right. This genome was sequenced with the very prevalent technique of paired-read sequencing, meaning the DNA fragments were sequenced on both ends (see Figure 9.18). Read 2 is paired with read 1 and is said to be the mate of read 1. Both reads are facing inward, and the middle part of the DNA fragment is left unsequenced. The paired reads are separated in the IGV reads track in order to maximize space. We can, however, request to view the reads as pairs by right-clicking the reads track and selecting View As Pairs.

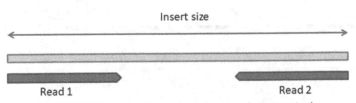

Figure 9.18: UCSC Genome Browser indicating that a particular DNA fragment was sequenced from both ends

The result in Figure 9.18 shows the reads pairs connected by a thin line, which corresponds to the part of the insert that was not sequenced.

If we click any read, a window pops up that provides the following information:

- **The length of the read:** The target read length is specified at sequencing and is usually 100 or 150 bp. In practice, the read length can vary slightly. Here, the vast majority of reads have a length of 101 bp.

- **The length of the insert:** This is quite variable as it is affected by the fluctuation in size of the original DNA fragments. The insert sizes vary here between 200 and 400 bp.

- **The mapping quality (MAPQ) of the read:** The mapping quality of a read equals $-10 \log 10$ (Probability{mapping position is incorrect}).

Conversely, probability{mapping position is incorrect} equals $10^{(-MAPQ/10)}$.

A tool called the Burrows-Wheeler Aligner (BWA), using an algorithm called Maximum Exact Matches (MEM), assigns a mapping quality (MAPQ) score that describes the probability that a sequences is mapped correctly to the reference genome. The maximum MAPQ score is 60, which corresponds to a probability of incorrect mapping of 0.0001 percent, which is one in a million (other MAPQ values and their corresponding percentages appear in Table 9.3). Reads with a very low MAPQ are displayed in IGV in white as opposed to their usual gray color.

Table 9.3: Mapping Quality (MAPQ) Scores and Their Corresponding Probabilities

MAPQ	PROBABILITY THE MAPPING POSITION OF A READ IS INCORRECT
0	100%
10	10%
20	1%
30	0.1%
40	0.01%
50	0.001%
60	0.0001%

Have a CIGAR

CIGAR stands for Concise Idiosyncratic Gapped Alignment Report. It is a compressed representation of an alignment that is part of the SAM file format. M stands for Match, D for deletion, I for insertion, S for soft clipping (part of a sequence that does not align to the reference but is still saved in the BAM file), and H for hard clipping (part of a sequence that does not align to the reference and is not saved in the BAM file). For example:

- **100M:** Perfect alignment.
- **80M5S15M:** The first 80 bp of the read perfectly match the reference, the next 5 bp are different from the reference, and the last 15 bp of the read again perfectly match the reference.

Paired reads with an insert size that is larger than expected are highlighted in red and can correspond to a deletion: the paired reads were close to each other in the DNA fragment of the individual but are separated by a big gap when mapped to the genome reference, as the individual is missing this area in their genome.

Conversely, paired reads with an insert size that is smaller than expected are highlighted in blue and can correspond to an insertion: the paired reads were reasonably far apart in the DNA fragment of the individual but were separated by a sequence that is unique to this individual and absent from the genome reference. As a consequence, the reads are very close to each other when mapped to the genome reference or can even overlap.

Figure 9.19 shows an isolated reads pair in a darker color with an insert size greater than 500 bp. This is not enough evidence to call a deletion event. In a true deletion event, several reads pairs are in a darker color around the same genomic location, and the coverage significantly drops as the aligner is struggling to map the reads around the deletion. Same rules apply for insertion events.

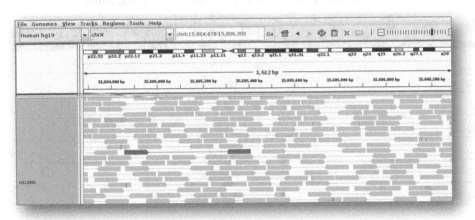

Figure 9.19: An indication that the aligner is having trouble mapping a sequence

As shown in Figure 9.19, the paired reads should be oriented inward. If the paired reads are both oriented toward the right or toward the left, this can correspond to an inversion. An *inversion* is a large section of DNA that is reversed in the sequenced individual compared to the reference genome. If the paired reads are both oriented outward, a duplication or translocation might have occurred in the sequenced individual compared to the reference genome.

Analyzing Variants in IGV

Let's go to the homeostatic iron regulator (HFE) gene by typing this gene name in the search box and clicking Go. Zoom in until the letters of the genome reference become visible. The HFE gene produces a protein that regulates iron levels in the body.

The sequence track displays the genome reference (hg19) the BAM was aligned to. If the sequence corresponds to an exon (which is the case in Figure 9.20), the corresponding protein sequence is displayed.

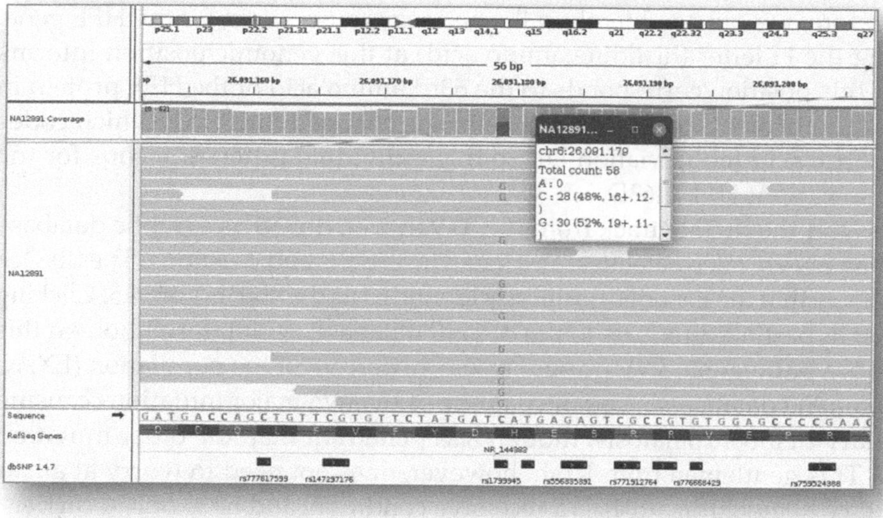

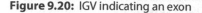

Figure 9.20: IGV indicating an exon

While the letters of the genome reference are visible, the letters of the aligned reads are visible only if they differ from the genome reference. We can see that, at position chr6:26,091,179, about half the aligned reads have the nucleotide G instead of C. These G letters have a strong hue and are not faded and therefore have good-quality scores. Clicking these individual bases displays their quality scores, which vary between 35 and 41.

The coverage track is colored at this genomic position. By default, the coverage track is highlighted in color when a nucleotide differs from the genome reference by more than 20 percent of the reads. This default can be modified by changing the "coverage allele-fraction threshold" in the Alignments tab of the Preferences menu. If you click this colored area in the coverage, a window pops up that summarizes the number of reads at that position, the total counts by letter at that position, and the strand (plus or minus) where these letters were identified.

We can see that 52 percent of the letters at that position are Gs instead of the reference C.

- There is no extreme strand bias (such as most guanines detected on the plus strand, and most cytosines detected on the minus strand), which could indicate an artifact.

- The genomic location has enough reads (depth) at this location.

- There is uniformity of coverage; there is no lower or higher depth versus neighboring regions, which could indicate misalignments.

- The reads mapping quality is good.

- The base quality scores of the guanines are excellent.

Therefore, we can confidently call a heterozygous variant for the HFE gene. Clicking the H letter (histidine amino acid) at this genomic position informs us that this position corresponds to the 63rd amino acid of the HFE protein in exon 2. The triplet CAT, coding for valine, is changed into GAT, which codes for aspartic acid (abbreviation D). So this individual is heterozygous for the missense variant HFE H63D.

Let's load the dbSNP track from the IGV server. dbSNP is a public database of human SNVs, microsatellites, and small insertions and deletions. The dbSNP track shows that the genomic position is identified as the SNP rs1799945. Clicking the SNP hyperlink directs us to the dbSNP database, with information on this SNV: it is a pathogenic SNV present in 7 to 10 percent of the population (EXAC and gnomAD databases). This SNV is one of the two major mutations causing hereditary hemochromatosis, though less penetrant than the other mutation C282Y. This gentleman from Utah, however, does not need to worry at all, as hereditary hemochromatosis is a recessive condition, and he is only a carrier.

We have looked at SNV. Insertions and deletions can also be visualized with IGV. Insertions are represented by a dark **I** for the insertion of one base. If more than one base is inserted, the number of inserted bases is displayed on top of the enlarged symbol. Figure 9.21 displays an example of insertions of four bases. Note that this insertion occurs just at the beginning of a series of TG repeats. When DNA polymerase encounters repeats during DNA replication, it can slip and cause an expansion of these repeats. An entry in dbSNP (rs10642017) features at this genomic location and describes insertion variants that occur at this locus with a frequency of 50 percent.

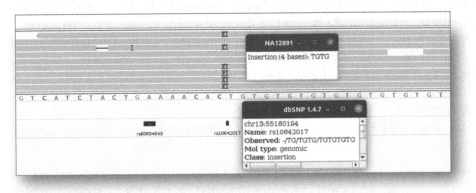

Figure 9.21: IGV indicating an inserted sequence

Deletions are highlighted by a break in the read, replaced by a black line.

Summary

This chapter discussed how to connect from your desktop to your AWS data and visualize and analyze remote BAM files. We identified different types of public annotations and genomic data that can be added as supplementary tracks to your data to help you interpret it. With IGV, we browsed through BAM files and made calls for genomic variants. Chapter 10 describes how to automate the discovery of genomic variants and their annotation.

10

Containerizing Your Workflow on the Desktop

This chapter will teach you about *containerization*, which is a way of stripping your applications down to their essentials and running only those key parts as an easily managed module—a container on a host computer. When you use flexible and scalable AWS services to run that host computer, you can have a very powerful genomics working environment to use at minimal cost.

Introducing Containerization

At one time, long ago, your only option for running applications was a physical server of some kind. Whether it was on your premises or hosted in a data center or some other way, there had to be a physical box with one or more physical processors, some memory, and some storage. This was an inefficient approach to the problem of providing applications. It was devilishly hard to strike a balance between making maximum use of expensive hardware (depreciating in value all the time) and having enough application capacity to handle surges in demand. One was either wasting capacity or annoying application users with shortfalls in performance and availability.

Virtual machines (VMs) provided an intermediate solution. With virtual machines, you could have images of applications stored as large files. These images contained all the information about the server that was required to run an application. They included an operating system, all the application software, and all of its dependencies. The image files could then be loaded by a virtual machine management application, the function of which was to share the physical resources of a computer across multiple virtual machines running on it. The virtual machine management application had responsibility for keeping track of memory regions across all of its hosted virtual machines and for sharing access to common resources, such as network connections.

With virtual machines, you still had to have hardware (local or hosted). The difference was that you could allocate limited physical hardware (running the virtual machines) differently across the population of virtual machines (running applications). It was possible to spin up more of one kind of virtual machine and fewer of another to take care of situations in which demand for one kind of application grew and demand for another ebbed.

The hardware, in a VM setup, is virtualized. A special software layer called a *hypervisor* manages the sharing of the hardware across all of the virtual machines running under it. Figure 10.1 is a graphical representation of several virtual machines, each sharing the host machine's hardware and with its own complete operating system, running on a single hardware platform.

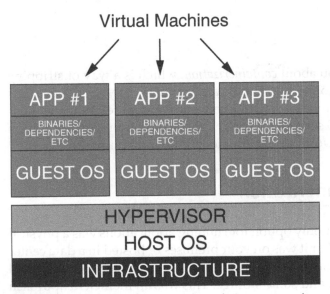

Figure 10.1: Virtual machines each have their own operating systems.

Amazon EC2 (and other, similar services) probably represent the ultimate evolution of the virtual machine idea. Not only does EC2 support images that can be stored as files (AMI images) and brought online as live virtual machines as required, but the whole infrastructure for hosting virtual machines is abstracted from the AWS user's point of view. Obviously, there are physical machines running EC2 instances in AWS data centers, but the average person who brings up an EC2 instance via the AWS console or command line is not (indeed, cannot be) concerned with how that works behind the scenes.

Because the hardware running EC2 instances is abstracted to the user, users can treat it as a resource to be drawn on as required. If your application needs more computing power, just request more—your budget will almost certainly run out before the supply of AWS computing power is exhausted. You can even set up your applications so that they scale automatically, bringing additional or larger EC2 instances online to run your application as dictated by demand.

EC2 has its drawbacks, though. For one thing, EC2 instances are still servers, which means they have to be managed. While AWS takes care of the infrastructure that runs the instances, you still have to concern yourself with everything inside the image, including the operating system. That means it is your responsibility to apply all patches related to security and everything else. Putting patches in place, and making sure they don't mess up the functionality of your application, is your time-consuming responsibility. Don't forget that operating systems sometimes have licensing costs.

As well, EC2 servers can still waste capacity. Even if your application is running on a small instance during a period of low demand, it is still racking up costs nonstop. The cost of running a t3a.small instance in us-west-1 (California) at this writing is on the order of USD $30 per month—not huge, but enough to add up over time.

What we want is something even more resource- and cost-efficient than scalable EC2 virtual machines. We want to define an application in its purest, simplest form and then run that minimal application in the most efficient way possible. We want to be able to put large amounts of computing power behind that application instance when it is required and put almost no resources into running it when the low demand allows for that.

This is the purpose of containers. *Containers*—there are many kinds, but Docker containers are the *de facto* standard—implement what is OS-level virtualization, which is to say virtualization of applications at the level of the operating system. One instance of an operating system can run many different containers. Applications running inside the containers "perceive" that they are running on their own computers, complete with processors, memory, peripherals, and storage. In fact, those physical resources (and the operating system on top of them) are potentially shared across many containerized applications.

Figure 10.2 shows a series of containers running as processes on a single operating system. The containers do not have their own operating system instances and do not share the host's hardware resources.

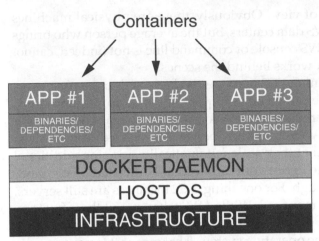

Figure 10.2: Containers have their own configurations on top of the host's operating system.

The following are the advantages of containerized architecture:

■ **Simplicity**: Once you have defined a Docker image—more on what is involved in that in the next section—you can pass it to someone else (or deploy it to a more capable execution environment) easily. This makes it easy to collaborate with others on the development of an image, and similarly easy to deploy a container described by an image locally for development, and into the AWS cloud when more computing power is needed.

■ **Flexibility**: You can do anything you want with a Docker image, unconstrained by hardware or other limitation.

■ **Comprehensiveness**: A container represents a closed universe. Everything that is required for a container to do what it is designed to do exists within that container. There is no need, when transferring an application from one computing environment to another, to make sure that dependencies are aligned, that there are no version conflicts, or that kind of thing. Give a container to someone else, or transfer it from your laptop to the AWS cloud, and you can expect it to do the same thing, with no surprises.

■ **Speed of start-up**: While it is not so much of an advantage in genomics analysis work, containers start up much faster than virtual machines. This means that in providing a service that has to respond to quickly evolving peaks and troughs in demand—think of a multiplayer, networked video game or something that has to react quickly to extreme changes in financial trading volumes—it is easy, and quick, to spin up containers on demand. Similarly, it is easy to take containers offline when demand for their services goes down.

A limitation of OS-level application virtualization is that a given operating system cannot host a container running another operating system. The containers must be set up to work with the operating system of the host machine on which they are running. While a Linux server running containers can handle containerized applications based on other Linux distributions, it is not possible to run containers that expect a Microsoft Windows environment.

Now, think about combining the advantages of containers (simplicity, ease of deployment, straightforward management) with the advantages of virtual servers (such as those hosted on AWS). That combination yields the ability to scale the resources of the host operating system up and down practically infinitely, while allocating those resources to multiple containerized applications in a way that matches their individual requirements. It would be possible to easily replicate a single container across multiple virtual host servers for redundancy. This is exactly what happens with the various AWS container management services, and this chapter will show you how to use them in your genomics applications.

Understanding and Using Docker

Docker is one of many implementations of OS-level virtualization (others include LXC and Solaris containers, as well as chroot "jails" that allowed processes to be given constrained environments as far back as Version 7 Unix in the late 1970s). It represents the *de facto* standard in this technology and is the basis for all AWS container management services.

A Docker *image* is a fixed file that defines a container, while a Docker *container* is a running instance of an image. You can think of the relationship between a Docker image and a Docker container as analogous to a Java class and an instance of that class. Figure 10.3 shows the relationship between a Docker image and containers that are spawned from it.

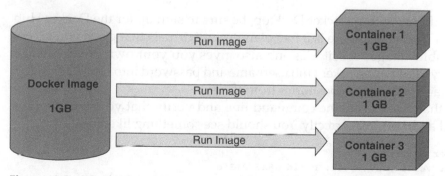

Figure 10.3: A Docker image can spawn many (potentially slightly different) containers.

Installing Docker on Your Local Machine

To work with Docker, you should have a copy of it installed on your local machine. You can install Docker by going to the Docker Hub (hub.docker.com), downloading the latest version, and following the installation instructions for your operating system. Once you have it installed, you'll need to run it and keep it running while you work with it at the command line. Docker comes with a monitoring application that allows you to keep an eye on containers you have running. It's shown (in its macOS version) in Figure 10.4.

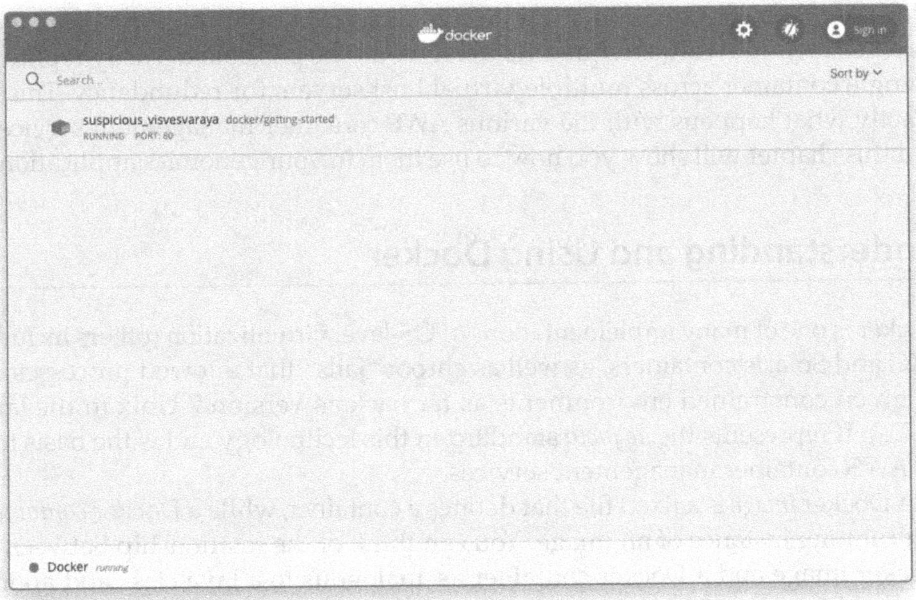

Figure 10.4: Docker showing a running container

When downloading Docker Desktop, be sure to sign up for the Docker Hub and get a Docker ID (there is a free plan for individuals that gives you access to the public image repositories and also gives you your own private image repository). Put your Docker Hub username and password into Docker Desktop for easy access to your repositories from there.

With that done, go to the command line and verify that you have Docker installed and running correctly. You should see something like this:

```
$docker --version
Docker version 19.03.12, build 48a66213fe
```

You can then run a simple Docker container ("Hello, world," as tradition demands) like this:

```
$docker run hello-world
Unable to find image 'hello-world:latest' locally
latest: Pulling from library/hello-world
0e03bdcc26d7: Pull complete
Digest: sha256:4cf9c47f86df71d48364001ede3a4fcd85ae80ce02ebad74156906
caff5378bc
Status: Downloaded newer image for hello-world:latest

Hello from Docker!
This message shows that your installation appears to be working
correctly.

To generate this message, Docker took the following steps:
 1. The Docker client contacted the Docker daemon.
 2. The Docker daemon pulled the "hello-world" image from the
Docker Hub.
    (amd64)
 3. The Docker daemon created a new container from that image which
runs the
    executable that produces the output you are currently reading.
 4. The Docker daemon streamed that output to the Docker client,
which sent it
    to your terminal.
```

You can see that Docker is smart enough to go and download images from its repository (the Docker Hub), searching for them by name, if they do not exist on your local machine. Let's take a look at some of the other features of Docker Desktop.

Downloading a Docker Image

A lot of the work you do with Docker will involve downloading images from the Docker Hub for use on your local machine. For illustration purposes, let's download an image that contains a Perl programming environment. The command for that is simple.

```
$ docker pull perl
Using default tag: latest
latest: Pulling from library/perl
57df1a1f1ad8: Pull complete
71e126169501: Pull complete
1af28a55c3f3: Pull complete
03f1c9932170: Pull complete
```

```
65b3db15f518: Pull complete
4c21f9be7406: Pull complete
c8090f4b73c1: Pull complete
Digest: sha256:a107c1e913309e35792b9e29caf29b15e44f64eab263e1f00beaa0
dc7ea26fdb
Status: Downloaded newer image for perl:latest
docker.io/library/perl:latest
```

That makes a Perl image available for running on the local machine.

Viewing Available Docker Images

To see which images we have available on the local machine, we can use `docker images`. It gives us a list of all the images we have built or downloaded, like this:

```
$ docker images
REPOSITORY              TAG      IMAGE ID      CREATED        SIZE
perl                    latest   e92f3fef8deb  2 weeks ago    859MB
docker/getting-started  latest   1f32459ef038  2 months ago   26.8MB
hello-world             latest   bf756fb1ae65  8 months ago   13.3kB
```

Running a Docker Container Interactively

To run a Docker container (to take an image and turn it into a running container, in other words), use this command:

```
docker run -it --rm perl
```

The `-it` option specifies that you want to run the container interactively (i.e., with a command line) and that there should be a terminal interface to allow you to use that command line.

Figure 10.5 shows the Docker UI updated with the newly started perl container running interactively.

Every container is different and can take its own commands interactively, but you can leave the interactive terminal associated with the running container by typing the following at the command line within the container:

```
exit
```

Figure 10.5: Docker showing a newly added container in operation

Removing a Docker Image

To get rid of a Docker container, use this command:

```
$ docker rmi perl
Untagged: perl:latest
Untagged: perl@sha256:a107c1e913309e35792b9e29caf29b15e44f64eab263e1f00beaa0
dc7ea26fdb
Deleted: sha256:e92f3fef8debbb1b831280695182dc0f3059d005151ac6f8660d9c1a5326744b
Deleted: sha256:8ca4c2018efade4ea0a49837e4bc0adae6bc2ae3003e7ce242f6bdeb89b9499b
Deleted: sha256:86d2b0158699dc9addb2edb2041ec7fe4be9db4ac44c5ab3392ed9cfb34a7cc5
Deleted: sha256:c54fd0b092afa958671934519fc3c33ea83096094e3683b63b27cbf92ece78a6
Deleted: sha256:e40886afa47f1a991f807ca03358366b86c1f65471c6c7e10187cf94eae5bcd1
Deleted: sha256:28189e70cc47eb3c32769ee0aad10e6b18a8886206db4df9ccfacb50878ca1ed
Deleted: sha256:92b912e56b356aeb8e2ab873bdf557ee80fbd3bec4482684fa0f0b0c30896acc
Deleted: sha256:4ef54afed7804a44fdeb8cf6562c2a1eb745dcaabd38b1ac60126f0966bf6aef
```

With that done, you can see that the image is gone.

```
$ docker images
REPOSITORY                   TAG        IMAGE ID        CREATED        SIZE
docker/getting-started       latest     1f32459ef038    2 months ago   26.8MB
hello-world                  latest     bf756fb1ae65    8 months ago   13.3kB
Lisieux:~ davidwall$
```

A Docker image is a recipe for creating a container. An image, rather than being a snapshot of a disk volume or other storage medium, consists of code that describes how to build a particular kind of container from the ground up. It might say to start with a particular base image from the Docker repository (more on that later), add some dependencies, and then add an application (taken from some other repository) on top of that. It is possible to think of images as consisting of layers: the configuration layer is on top of the application software layer, which is on top of a dependency layer, which is in turn on top of an operating system layer.

A container, then, is a running instance of a Docker image. If you know something about object-oriented software architectures, you might think of an image as an object class and a container as an object created from such a class.

More on Using the Docker Hub

Central to the Docker experience is the Docker Hub, which is a set of repositories (public and private) in which Docker images are stored. You can access the Docker Hub via a browser (at `hub.docker.com` using your Docker username and password to log in). If you prefer, you can get there via a menu option in the Docker Desktop application.

The Docker Hub is searchable. Say you wanted a container that ran a database. You could put the name of the database into the search box or browse for databases by category. The Docker Hub will bring up a list of available images that match. Figure 10.6 shows the listing for MySQL, the popular open-source structured database server.

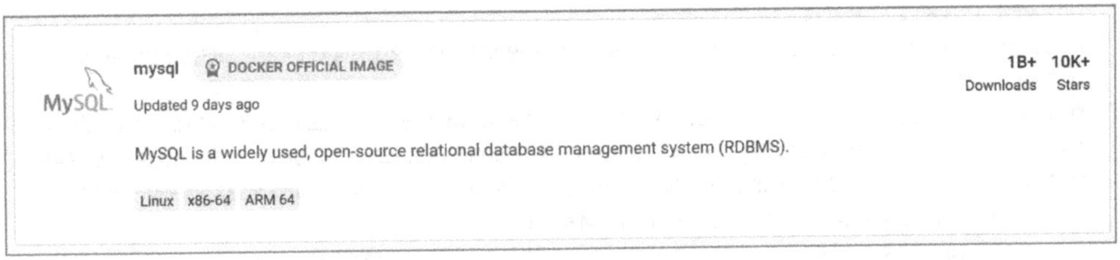

			1B+	10K+
MySQL	mysql ♀ DOCKER OFFICIAL IMAGE		Downloads	Stars
	Updated 9 days ago			
	MySQL is a widely used, open-source relational database management system (RDBMS).			
	Linux x86-64 ARM 64			

Figure 10.6: A Docker image for running the MySQL database

Note the "Docker Official Image" tag in the listing. It indicates that this image is maintained by people who know what they are doing—either from the company that owns the software or from the open-source community that maintains it. You can generally trust images tagged as official to perform correctly and without surprises.

Containers for Genomics Work

For genomics work, a number of prebuilt Docker images exist. For the purposes of illustration here, we will download and run the Genomes in the Cloud image, designed and maintained by the Broad Institute. In the Docker Hub, search for the following:

```
broadinstitute/genomes-in-the-cloud
```

You will get a matching Docker image—specifically, the official Broad Institute Genomes in the Cloud (GITC) image. Figure 10.7 shows the results.

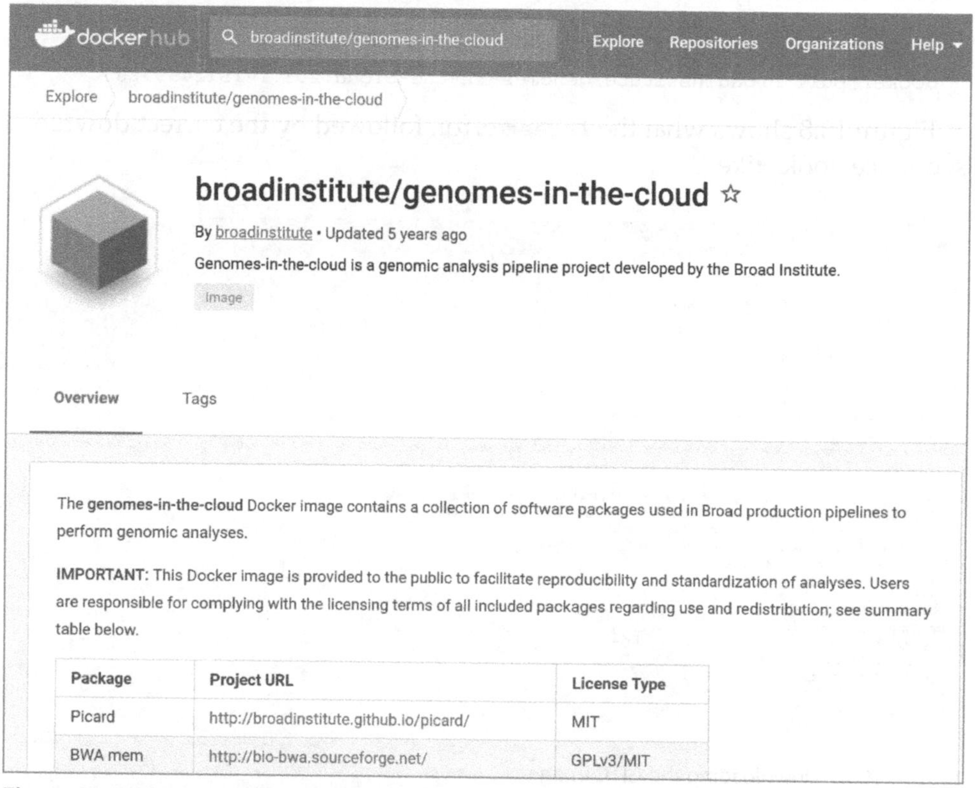

Figure 10.7: The Broad Institute's Genomes in the Cloud image

The first one in the list is the image we want. By clicking it, we get a page that confirms that the image includes a lot of the tools we will want to use in our genomics work, including the following:

- Picard, a set of command-line tools for manipulating genomics files
- BWA, an implementation of the Burrows-Wheeler Aligner for comparing experimental samples with a reference genome
- GATK, the widely used Genome Analysis Toolkit developed and maintained by the Broad Institute
- Various utilities and languages, such as Java, Python, and R

The `pull` command appears on the right side of the page, complete with a clickable link to copy the command into your local clipboard.

Now, pasting that command at a command line and running it, like so:

```
docker pull broadinstitute/genomes-in-the-cloud
```

ordinarily would cause the image to be pulled down from the Docker Hub repository, using the default `latest` tag that usually denotes the most recent version of an image. However, the GITC image does not have a `latest` tag, so we must manually specify the latest version as a tag, like this:

```
docker pull broadinstitute/genomes-in-the-cloud:2.3.1-1512499786
```

Figure 10.8 shows what the `latest` error, followed by the correct download sequence, looks like.

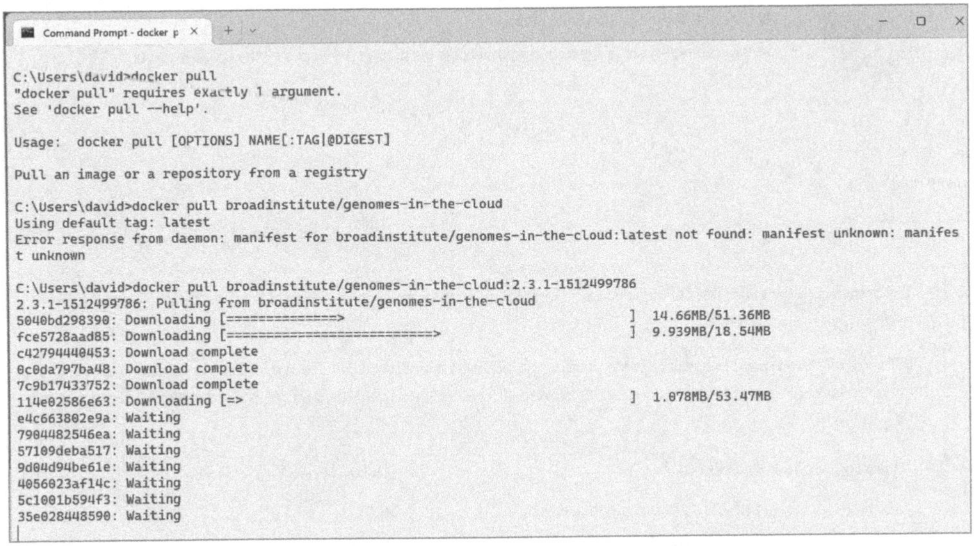

```
C:\Users\david>docker pull
"docker pull" requires exactly 1 argument.
See 'docker pull --help'.

Usage:  docker pull [OPTIONS] NAME[:TAG|@DIGEST]

Pull an image or a repository from a registry

C:\Users\david>docker pull broadinstitute/genomes-in-the-cloud
Using default tag: latest
Error response from daemon: manifest for broadinstitute/genomes-in-the-cloud:latest not found: manifest unknown: manifes
t unknown

C:\Users\david>docker pull broadinstitute/genomes-in-the-cloud:2.3.1-1512499786
2.3.1-1512499786: Pulling from broadinstitute/genomes-in-the-cloud
5040bd298390: Downloading [==============>                      ]  14.66MB/51.36MB
fce5728aad85: Downloading [===========================>          ]  9.939MB/18.54MB
c4279440453: Download complete
8c0da797ba48: Download complete
7c9b17433752: Download complete
114e02586e63: Downloading [=>                                    ]  1.078MB/53.47MB
e4c663802e9a: Waiting
7904482546ea: Waiting
57109deba517: Waiting
9d04d94be61e: Waiting
4056023af14c: Waiting
5c1001b594f3: Waiting
35e028448590: Waiting
```

Figure 10.8: Downloading the GITC image

With a GITC image downloaded, we can run it.

```
docker run -it --rm  broadinstitute/genomes-in-the-cloud:2.3.1-1512499786
```

Within moments, we have a running GITC container and find ourselves at its command line, within our local console.

```
Lisieux:~ davidwall$ docker run -it --rm  broadinstitute/genomes-in-the-cloud:2.3.1-1512499786
root@fccdbdee0d48:/usr/gitc# ls
Dockerfile  GATK35.jar    bgzip  gatk4        svtoolkit2.00
GATK34.jar  GATK36.jar    bwa    picard.jar   tabix
```

A quick sanity check proves that we can run GATK, as shown here:

```
root@fccdbdee0d48:/usr/gitc/gatk4# ./gatk-launch --help

 Usage template for all tools (uses --sparkRunner LOCAL when used with a
Spark tool)
    ./gatk-launch AnyTool toolArgs

 Usage template for Spark tools (will NOT work on non-Spark tools)
    ./gatk-launch SparkTool toolArgs  [ -- --sparkRunner <LOCAL | SPARK | GCS>
sparkArgs ]

 Getting help
    ./gatk-launch --list        Print the list of available tools

    ./gatk-launch Tool --help  Print help on a particular tool

 gatk-launch forwards commands to GATK and adds some sugar for submitting
spark jobs

    --sparkRunner <target>      controls how spark tools are run
      valid targets are:
      LOCAL:      run using the in-memory spark runner
      SPARK:      run using spark-submit on an existing cluster
                  --sparkMaster must be specified
                  --sparkSubmitCommand may be specified to control the Spark
submit command
                  arguments to spark-submit may optionally be specified after --
      GCS:        run using Google cloud dataproc
                  commands after the -- will be passed to dataproc
                  --cluster <your-cluster> must be specified after the --
                  spark properties and some common spark-submit parameters will
be translated
                  to dataproc equivalents

    --dryRun       may be specified to output the generated command line without
running it
    --javaOptions 'OPTION1[ OPTION2=Y ... ]'   optional - pass the given string
of options to the
                  java JVM at runtime.
                  Java options MUST be passed inside a single string with space-
separated values.
```

As well, we can run the following to see all the tools and options available within GATK:

```
./gatk-launch -list
```

We have a working genomics toolkit running as a container on our local machine.

To get out of the container's command line and return to the command line of the host computer (your laptop or whatever you're using), use the `exit` command.

You may find that the `exit` command does not work and that the container instead complains about stopped jobs. You can see all the jobs you have running with the `jobs` command.

```
root@fccdbdee0d48:/usr/gitc/gatk4# jobs
[1]+  Stopped                 cat
```

In this case, this shows that an instance of the file-listing command, `cat`, is still present in Stopped state. You can get its process ID (pid) with the `ps` command.

```
root@fccdbdee0d48:/usr/gitc/gatk4# ps
  PID TTY          TIME CMD
    1 pts/0    00:00:00 bash
   20 pts/0    00:00:00 cat
   49 pts/0    00:00:00 ps
```

In this case, `ps` returns 20 for the misbehaving `cat` process. (Process IDs are typically much smaller in Docker containers than in full Linux systems.)

It can't be killed in the usual way, despite no error being thrown.

```
root@fccdbdee0d48:/usr/gitc/gatk4# kill 20
```

That is because of a quirk in the way process IDs work in containers. Without getting into it, the easy way to get rid of a hanging process is to use a stronger variant of the `kill` command, like this:

```
root@fccdbdee0d48:/usr/gitc/gatk4# kill -9 20
[1]+  Killed                  cat
```

Summary

This chapter discussed containers and their relative merits with respect to virtual machines. You saw that containers are much leaner and faster to start than VMs, which is great for handling workloads in which demand goes up and down rapidly—and also excellent for genomics researchers who want to be able to develop workflows on their local machines before uploading the containers to more capable computing resources, such as those in the AWS cloud.

That is the subject of our next chapter. We will show you how to take container images that you have developed on your local machine, store them in an AWS container repository, and then run those containers securely and cost-effectively using AWS container services.

Variants and Applications

In this chapter, we look at some applications of the analysis of genomes. We first investigate polygenic risk scores, which capture the additive effects of thousands of variants to influence our complex traits. We then extend our study to metagenomics and explore how metagenomics can be applied to human samples.

Polygenic Risk Scores

We know from twins studies that certain traits and health conditions are highly hereditary. Studies of identical and fraternal twins suggest, for example, that up to 80 percent of height variation is genetic. However, as the determination of height involves many variants along the genome, up to recently only 5 percent of height heritability could be explained. This paradox is referred to as the *missing heritability problem*. Rare diseases are often caused by a defective gene; in contrast, many traits and most common diseases are caused by hundreds or thousands of genetic variations.

Genome-wide Association Studies

Finding these causal variants requires undertaking large studies called *genome-wide association studies* (GWASs), which look for associations between traits and genetic variants, in other terms, associations between phenotype and genotype, in a population. Although GWASs can study copy number variants or other sequence variations, most GWASs focus on single nucleotide variants (SNVs). GWASs were performed from microarrays in the past and are currently commonly conducted from WXS or WGS. There are about one million independent common genetic variants along the genome. So a lot of statistical tests must be performed, which must be corrected for multiple hypothesis testing. To achieve statistical significance, GWASs need therefore to be conducted on large cohorts, sometimes hundreds of thousands of people, or aggregate a large number of studies (*metastudies*).

Initial GWAS analyses studying height identified 40 genetic variants that could explain 5 percent of height variation. In 2018, a global consortium called GIANT performed a GWAS on 700,000 people that found 3,300 genetic variants accounting for 25 percent of height variation. In 2020, the same consortium aggregated DNA information from 4.1 million individuals to identify 9,900 variants that explain 40 percent of the variation in height. While the height heredity is still not fully accounted for, this endeavor highlights the incredible complexity of many traits and of most common diseases, as well as the multiplicity of causal variants with each a very small effect.

More than 5,700 GWASs have been performed so far, analyzing more than 3,000 traits. GWASs have historically focused on populations of European ancestry, and the findings are less applicable to other ethnicities. However, a growing number of GWASs now focus on other ethnicities.

Interestingly, while GWASs sometimes identified genetic variants in genes or regulatory areas, the majority are located in noncoding regions. This is quite humbling. We know that DNA can fold into loops to bring remote regions of the genome in contact and allow these remote regions to act as enhancers and regulate the transcription of genes located far away. However, we lack a mechanism to explain many of the GWAS-identified noncoding variants.

Some SNVs are known to affect the expression of mRNA and are referred to as *eQTL*, which stands for "expression quantitative trait loci." Many studies have used eQTL analysis to explain the results of GWASs. Comprehensive catalogs of eQTL are publicly available for many tissue types (remember, mRNA expression is tissue-specific). The Genotype-Tissue Expression (GTEx) is the largest resource with eQTL cataloged for 49 tissues; it is available at gtexportal.org. The eQTLGen Consortium identified the eQTL derived from the blood samples of about 30,000 individuals (www.eqtlgen.org).

GWASs have attracted criticism as they implicate so many genetic variants for common diseases that it appears difficult to use the findings to make treatments to help patients. GWASs helped discover some biological mechanisms behind certain conditions, such as autophagy in Crohn's disease and the complement pathway in subtypes of macular degeneration and schizophrenia, though it has not led to new treatments. The genetic variants uncovered by GWASs are, however, widely used to help predict traits and risks of common diseases using polygenic risk scores.

Calculating a Polygenic Score

Polygenic risk scores (PRSs), also called *polygenic scores* (PGSs) or *genomic scores*, predict phenotypes, such as a trait or disease. How are they calculated? The results of GWASs consist of a list of genetic variants (alleles at particular loci on the genome) with an effect size for each variant. The polygenic risk score is calculated by computing the sum of risk alleles that a person has, multiplied by their corresponding effect sizes as estimated by a GWAS on the phenotype.

$$PRS = \sum_{i=0}^{N} \beta_i * \text{dosage}_i$$

N is the number of genetic variants or SNVs in the score, β_i is the effect size (also called *beta*) of the variant *i*, and dosage *i* is the number of copies of SNV *i* in the studied person.

A catalog of polygenic risk scores is available at `www.pgscatalog.org` and currently contains more than 2,600 polygenic scores for nearly 550 traits.

Are the polygenic risk scores accurate? We saw that we can explain only half of the heritability of height, though it has been the focus of many large studies. In addition, complex diseases are the result of a combination of genotype and environmental factors. So they are indicative and work best when combined whenever possible with environmental factors.

Polygenic scores can be calculated by software dedicated to calculating such scores, such as PRSice, LDpred, JAMPred, PRS-CS, or lassosum, or generic tools for genomic analysis such as the popular PLINK or biqsnpr. We will use PLINK 2.0 here.

Let's install PLINK 2.0.

```
> wget https://s3.amazonaws.com/plink2-assets/alpha3/plink2_linux_avx2_
20220603.zip
> unzip alpha3/plink2_linux_avx2_20220603.zip
```

Select a PRS of interest from the PGS Catalog or from the supplementary information of a publication. Carefully consider the units used to measure the phenotype of interest, the confounding factors, the ethnicity of the cohort, and the predictive capability of the PRS. PRSs are commonly evaluated using the pseudo-R^2, which corresponds to the amount of phenotypic variance explained by the PRS.

The input information to calculate a PRS should include the following:

- The list of SNVs composing the score. They can be specified with an rsID or RefSNP or with the bp position in the genome.
- The effect allele for each SNV.
- The effect size (beta or weight) for each SNV.
- The genome reference.

If the effect size is specified as a hazard ratio or an odds ratio, take the log of the ratio to obtain the weight.

As an example, Table 11.1 shows an extract of a PGS Catalog scoring file from a recent GWAS on thyroid disease. The study identified nearly 900,000 associated variants, with small effect sizes.

Table 11.1 Example of PGS Catalog Scoring File

```
#genome_build=GRCh37
#variants_number=895602
#weight_type=BETA
```

CHR_NAME	CHR_POSITION	EFFECT_ALLELE	OTHER_ALLELE	EFFECT_WEIGHT
1	754182	A	G	-2.0E-06
1	768448	G	A	-2.0E-06
1	779322	A	G	8.0E-06
1	846808	C	T	2.0E-06
1	853954	C	A	-1.3E-05
1	854250	A	G	3.0E-06
1	861808	A	G	1.7E-05
1	863124	G	T	3.9E-05

The SNVs of the genomes to analyze can be extracted by utilities such as BGEN, which is directly supported by PLINK, or bgenix. The extracted SNVs then need to be converted to PLINK format.

```
> plink2 --bgen single_allelic_file.bgen ref-first \
--hard-call-threshold 0.1 \
--sample mytestsample.sample \
--memory 16000 \
--set-all-var-ids @:#_\$r_\$a \
--freq \
--make-pgen \
--out raw
```

The input single allelic file includes only the alleles specified in the PGS scoring file. We used the default threshold of 0.1 for hard calls.

The output is a set of PLINK files.

- **File with extension pgen:** Individual identifiers (IDs) and genotypes
- **File with extension pvar:** Information on SNVs
- **File with extension psam:** Information on the individuals
- **File with extension afreq:** Allele frequencies

To ensure good practices, PLINK requires a SNV quality control file and a sample quality control file to be able to calculate the PRS score.

Let's perform quality control on the SNVs and generate the snvQC file. We exclude the ambiguous SNVs and the SNVs with less than 0.3 imputation information.

```
> plink2 --pfile raw \
--memory 16000 \
--exclude exclrsIDs_ambiguous.txt \
--extract-col-cond test.tsv 9 10 \
--extract-col-cond-min 0.3 \
--maf 0.005 \
--write-snplist \
--make-pgen \
--out snvQC
```

Similarly, we perform quality control on the sample to analyze, by only keeping samples belonging to a list of individuals (`samples-that-passed-QC.txt`) that passed criteria such as no sex discordance or unrelated individuals.

```
> plink2 --pfile raw \
--memory 16000 \
--extract snvQC.snplist \
--keep-fam samples-that-passed-QC.txt \
--write-samples \
--out sampleQC
```

We are now able to calculate the PRS, which is computed as follows:

$$PRS = \sum_{i=0}^{N} \frac{\beta_i * \mathrm{dosage}_i}{P * M}$$

where P is the ploidy (2 for humans) and M is the number of nonmissing SNVs in the studied individual.

```
plink2 --pfile raw \
--memory 16000 \
--extract snvQC.snplist \
```

```
--keep sampleQC.id \
--score score.txt no-mean-imputation \
--out PRS
```

Metagenomics

Metagenomics refers to the study of genomes isolated from bulk samples, typically from environmental samples from the water, sea, or soil. These samples contain ecosystems of microorganisms that can be sequenced in bulk.

Metagenomics can also apply to human samples. Communities of microorganisms including bacteria, viruses, and fungi live on our skin, in our mouth, and in in our intestinal tract. Even our blood, which used to be seen as sterile, harbors viruses as well as bacteria that have translocated from our guts or lungs. When we sequence a human sample, whether a blood or saliva sample or a cancer tumor, all DNA present in the sample gets sequenced. All reads get mapped to the human genome reference, and the unmapped reads are discarded. These unmapped reads actually correspond to other organisms that were present in the sample at the time of sequencing and therefore were sequenced as well.

Do all these reads correspond to organisms that were actually present in the individual at the time of the sample taking? No. The majority of unmapped reads map to water bacteria that are present in the laboratory reagents. Some reads correspond to artificial viral vectors that are commonly used in laboratories to transfect cell lines with genetic material. So metagenomics studies require strict quality control to detect these contaminations early in the processing pipelines. However, the study of unmapped reads can be very informative and detect microorganisms that contribute to disease.

Metagenomics studies of bacteria and fungi usually involve specialized software that relies on a large database of all sequenced bacteria and fungi and has algorithms to match sequenced reads with organisms. Viruses can be studied the same way. However, as viral genomes are very small compared to bacteria and fungi, viruses can be studied by building a concatenated viral reference of all viruses ever sequenced and mapping the unmapped reads to this viral reference.

Let's first create the viral reference. After downloading all viral genomes from NCBI at www.ncbi.nlm.nih.gov/assembly, we can concatenate these genomes into a single viral reference.

```
>cat * > viral_reference.fa
```

We use the samtools command to extract the unmapped reads from the sample to analyze.

```
samtools view -f 0x04 -h -b original.bam -o unmapped.bam
```

To realign the extracted reads to the new viral reference, it is necessary to convert the unmapped reads from BAM to FASTQ format. The BAM file needs first to be sorted by read name to keep the records in the two FASTQ files in the same order.

```
samtools sort -n -o unmapped.qsort.bam unmapped .bam
```

```
bedtools bamtofastq -i unmapped.qsort.bam \ -fq unmapped.read1.fq \ -fq2
unmapped.read2.fq
```

Let's index the reference before realigning.

```
bwa index viralreference.fa
```

We can now align the unmapped reads to the viral reference.

```
bwa mem viralreference.fa unmapped.read1.fq unmapped.read2.fq > aln-viral.sam
```

This allows us to obtain the list of detected viruses on the sample. This first list needs to be filtered to select only those reads above a certain mapping quality (a MAPQ greater than 30, for example) and with the aligned read matching a minimum number of bp of the viral sequence. As viral sequences are often rearranged, it is better to allow the minimum number of matching base pairs (80, for example) to not necessarily be consecutive. The following script performs this filtering:

```
cat foundviruses.txt | while read VIRUS ; do
for f in all-viral.sam;
do
samtools -q 30 -F 1280 $f $VIRUS |
awk -v outfile=${f##*/} ' function isnum(x){return(x==x+0)}
{beginfield = 1; newfield = 0; Mcount = 0; str
```

AlphaFold

AlphaFold has revolutionized the determination of the three-dimensional structure of proteins, which was previously done by homology modeling. AlphaFold gets better results by making use of deep learning, which is to say it identifies patterns by looking at proteins whose structure has already been determined (typically by crystallography). All the crystallography information is in a database called the *Protein Data Bank* (PDB), which has the coordinates of all atoms of pretty much every protein ever crystallized. It is big! So if you want to determine the 3D structure of a protein once in a while, the AlphaFold online server is enough. But if you want to do it relatively often, you can do it yourself on AWS.

Given the amino acid sequence of a protein, AlphaFold predicts the 3D structure of that protein, of how the string of amino acids folds up to become the functional protein structure. Looking under the hood of a car reveals a lot about how the engine makes the car carry out its function of transportation. Looking at the structure of a protein reveals a lot about how that protein carries out its function, whether that function be that of an enzyme creating or breaking molecules or a channel or transporter allowing molecules to cross the cell membrane into and out of the cell or a signaling protein passing on messages to other parts of the cell or a structural protein providing scaffolding for the cell. The structures of the normal versions—the wild-type versions—of all human proteins have now been predicted by AlphaFold and are available in the PDB. Go to www .uniprot.org, search by the name of the protein of interest, and after choosing the human version of the protein, click the Structure tab where the 3D structure predicted by AlphaFold will be available for viewing in the browser. When DNA sequencing reveals that there is a mutation in a coding region of a gene, one can run the AlphaFold software on the gene's protein sequence containing the mutated amino acid to see whether and how the protein's 3D structure has been changed by the mutation.

Predicting Protein Structure from Protein Sequence—A 50-Year Puzzle

Why is predicting 3D protein structure such a big deal? When it comes to proteins, structure is function, providing knowledge and understanding of how the protein works and, by extension, how the cell and how biology work. Knowing the protein structure allows the deliberate design of drugs to interact with the protein. Is the protein an oncogene that is aberrantly switched on in cancer, continually signaling to the cell to proliferate and grow the cancer tumor? Drugs have been designed to block oncogenes so that they don't work anymore and stop telling the cancer to grow, like jamming a lock with a defective key so that the real key can't be inserted. For example, the anticancer drug erlotinib is a molecule designed to fit snugly into an oncogene (the growth receptor called *EGFR receptor*) that is often switched on in lung cancers, blocking the oncogene and thus stopping it from signaling to the cancer cell to grow and proliferate. The design and success of this anticancer molecule was made possible by knowing the 3D protein structure of the oncogene. Now that AlphaFold is available, it is turbo-charging synthetic biology activities to deliberately design protein enzymes to carry out handy new functions, including biosensors and molecular motors.

Given the priceless knowledge contained in 3D protein structures, a lot of work has gone into working out the 3D structures of many proteins over the last 50 years. To solve the structure of a protein, a sufficient quantity of the protein had to be extracted and purified. Sometimes this involved putting on a winter coat and going into the cold room to work on a delicate protein that would fall apart at room temperature. When molecular biology and biotechnology techniques became available, they were used to coax bacteria or other cells in the laboratory to produce sufficient quantities of the protein, which then still had to go through the tedious and sometimes fickle purification processes. The most informative technique during the last 50 years for determining protein 3D structure has been X-ray crystallography. When each protein molecule is in the same orientation as every other protein molecule, that is, when the protein is crystallized, then firing an X-ray beam at the crystal will cause the X-rays to be deflected by the crystal in an organized way. The pattern image that the X-rays make on the detector can be interpreted to determine the position of each atom in the protein in the crystal! Most proteins don't naturally crystallize, which is a good thing for us because we don't want the proteins in our body to turn into crystals. This means a lot of trial and error is required to find the conditions under which a given protein will crystallize. For many proteins, particularly for membrane proteins, it wasn't possible to produce crystals.

Another technique used for solving protein structures is nuclear magnetic resonance (NMR), using similar technology to the magnetic resonance imaging (MRI) used to take medical images of patients. Recently, cryogenic electron microscopy (cryo-EM) has come of age and is now capable of determining protein structures to higher resolution than even X-ray crystallography, including difficult-to-solve membrane proteins and protein-complexes. Cryo-EM involves taking thousands of images of protein molecules by electron microscopy and using high-performance computing, which has only recently become available, to analyze the images. Cryo-EM still requires the long, hard work of protein extraction and purification. Determining one protein structure could easily require 4 years of work of a PhD student, and sometimes these efforts fail to successfully determine the protein structure. By the year 2020, the result of this enormous amount of scientific labor was around 100,000 unique structures of proteins in the Protein Data Bank (PDB; www.rcsb.org), freely available to anyone and everyone, including you.

The string of amino acids composing the protein folds up into the 3D structure naturally, or with a bit of help from chaperone proteins, so that positive and negative charges of amino acids will be near each other and hydrophobic patches will be together. The protein 3D structure is encoded in the protein amino acid sequence. If only we knew how to interpret that code, we could determine the 3D structure simply by knowing the sequence. This is the protein-folding problem that scientists have been working on for the last 50 years. AlphaFold has now solved the protein-folding problem! To do this, AlphaFold used the artificial intelligence technique of machine learning. Its deep learning algorithm studied those 100,000 sequences and structures in the PDB to come up with the solution. The AlphaFold algorithm could not have done this if those 50 years' worth of experimentally determined structures were not available. Now we can computationally determine a protein's 3D structure in a day, instead of 4 years! However, we don't expect science to rest on its laurels. The experiments with cryo-EM and other techniques continue unabated, and the new knowledge generated will be used to create future AlphaFold models to predict large protein complexes and interactions between proteins and lipid membranes and drugs. After its successes with predicting structure for a single "monomer" sequence, AlphaFold has already come out with a "multimer" version to predict the structure of a protein complex composed of two or more amino acid sequence chains.

Installing and Running AlphaFold

The easiest way to start running AlphaFold is to run it for free in the Google Cloud in a Python Colab notebook specifically set up to make AlphaFold available to everyone, at `colab.research.google.com/github/sokrypton/ColabFold/blob/main/AlphaFold2.ipynb` or `colab.research.google.com/github/deepmind/alphafold/blob/main/notebooks/AlphaFold.ipynb`. However, given that these resources are provided for free, they may time out and not run to completion, and they are limited in how big a protein and how many proteins you can predict at a time. The most reliable way to run AlphaFold is to install and run it on resources that you control and probably pay for. AWS has provided instructions at `aws.amazon.com/blogs/machine-learning/run-alphafold-v2-0-on-amazon-ec2` on how to install and run AlphaFold in the AWS Cloud, and it is recommended to follow those instructions for running AlphaFold on AWS. Here are the explanations for those instructions.

AlphaFold uses machine learning, which requires substantial computing resources. AlphaFold has been optimized to efficiently leverage currently available high-performance computing resources such as GPUs developed for gaming. The first step involves choosing an appropriate AWS machine image that supports these machine learning optimizations, such as the Deep Learning Amazon Machine Image (DLAMI). The next step is to download the AlphaFold code from `https://github.com/deepmind/alphafold.git` and any dependent software that is not already pre-installed on the Amazon machine. AlphaFold needs a lot of the existing protein sequence and protein structure data, so the next step is to download that data. It could take a day to download and will take up nearly 3 TB of disk space (remember, it contains 50 years' worth of scientific results!). AlphaFold contains a lot of components, and the easiest way to install them is via the handy Docker container provided with the AlphaFold code. The AWS instructions explain how to build and install the AlphaFold container, including how to point the application to the protein data that you have downloaded. The AWS instructions also recommend how to create a snapshot of your AlphaFold installation so that it can be stored without the cost of running a machine when you aren't running AlphaFold, and quickly restored when you do want to run AlphaFold.

Here is the command to run AlphaFold using the AWS installation's Docker container (wrapped to two lines because of the page width):

```
nohup python3 /data/alphafold/docker/run_docker.py --fasta_paths=/data/
    input/myProteinSequence.fasta --max_template_date=2020-05-14 &
```

The input protein sequence for which you want to predict the 3D structure is in the `myProteinSequence.fasta` file, in the FASTA format. (The first line contains a > and a text identifier or description, and subsequent lines contain the amino acid sequence.)

Singularity is another container system and is available on AWS. If you prefer to run AlphaFold as a singularity container instead of as a Docker container, then here are the instructions for converting the Docker container to a singularity container:

```
sudo docker build -f docker/Dockerfile -t alphafold220 .
```

The 220 is for the version 2.2.0 available at the time of this writing.

At the end you will see something similar to the following. Note the build identifier.

```
Successfully built 126f1e5a412c
Successfully tagged alphafold220:latest
```

In this next command, use the build identifier that you noted.

```
sudo docker save 126f1e5a412c -o alphafold220_docker.tar
sudo singularity build alphafold220.sif docker-archive://alphafold220_
    docker.tar
```

If you are installing AlphaFold version 2.2.0 and find that you have error messages about incorrect or missing packages, then you may need to make the following changes to the AlphaFold/Dockerfile file and rerun the installation from the `sudo docker build` command.

```
nano alphafold/Dockerfile
```

Add this line as the first `RUN apt-key` line:

```
RUN apt-key adv --fetch-keys https://developer.download.nvidia.cn/
    compute/cuda/repos/ubuntu2004/x86_64/3bf863cc.pub
```

On and after the last "Install pip packages" section, add this:

```
RUN pip3 install --upgrade protobuf==3.20.0
### Then add these two lines after all installs and before the command:
  RUN chmod u+s /sbin/ldconfig.real
```

After all installs, make sure we have the correct protobuf version in case other installs have upgraded protobuf.

```
RUN pip3 install --upgrade protobuf==3.20.0
```

To run this AlphaFold singularity container for a monomer (only one) protein sequence, use this:

```
DATASET=/my/downloaded/alphafold/datasets
PROJDIR=/my/project/directory # with the input fasta file in input/
myMonomerSequence.fasta
singularity exec --nv        \
    --env TF_FORCE_UNIFIED_MEMORY=1,XLA_PYTHON_CLIENT_MEM_FRACTION=4.0 \
    -B $DATASET/:$DATASET/   \
    -B $PROJDIR:$PROJDIR  \
    alphafold220.sif     \
        /app/run_alphafold.sh        --data_dir=$DATASET    \
          --uniref90_database_path=$DATASET/uniref90/uniref90.fasta  \
          --mgnify_database_path=$DATASET/mgnify/mgy_clusters_2018_12.fa  \
          --pdb70_database_path=$DATASET/pdb70/pdb70  \
          --template_mmcif_dir=$DATASET/pdb_mmcif/pdb_mmcif/mmcif_files/  \
          --obsolete_pdbs_path=$DATASET/pdb_mmcif/pdb_mmcif/obsolete.dat  \
          --bfd_database_path=$DATASET/bfd/bfd_metaclust_clu_complete_
id30_c90_final_seq.sorted_opt  \
          --uniclust30_database_path=$DATASET/uniclust30/uniclust30_2018_08/
uniclust30_2018_08        \
          --fasta_paths=$PROJDIR/input/myMonomerSequence.fasta    \
          --max_template_date=2020-05-14        \
          --model_preset=monomer        \
          --db_preset=full_dbs        \
          --output_dir=$PROJDIR/output        \
          --use_gpu_relax=True        \
          -v 1
```

To run this AlphaFold singularity container for multimer (more than one protein sequence forming a protein complex), use this:

```
DATASET=/my/downloaded/alphafold/datasets
PROJDIR=/my/project/directory # with the input fasta file in input/
myMultimerSequences.fasta
singularity exec --nv        \
    --env TF_FORCE_UNIFIED_MEMORY=1,XLA_PYTHON_CLIENT_MEM_FRACTION=4.0 \
    -B $DATASET/:$DATASET/   \
    -B $PROJDIR:$PROJDIR  \
    alphafold220.sif     \
        /app/run_alphafold.sh        --data_dir=$DATASET    \
          --uniref90_database_path=$DATASET/uniref90/uniref90.fasta   \
          --mgnify_database_path=$DATASET/mgnify/mgy_clusters_2018_12.fa    \
          --template_mmcif_dir=$DATASET/pdb_mmcif/pdb_mmcif/mmcif_files/  \
```

```
        --obsolete_pdbs_path=$DATASET/pdb_mmcif/pdb_mmcif/obsolete.dat   \
        --bfd_database_path=$DATASET/bfd/bfd_metaclust_clu_complete_
id30_c90_final_seq.sorted_opt   \
        --uniclust30_database_path=$DATASET/uniclust30/uniclust30_2018_08/
uniclust30_2018_08    \
        --pdb_seqres_database_path=$DATASET/pdb_seqres/pdb_seqres/
pdb_seqres.txt \
        --uniprot_database_path=$DATASET/uniprot/uniprot/uniprot.fasta \
        --fasta_paths=$PROJDIR/input/myMultimerSequences.fasta    \
        --max_template_date=2020-05-14           \
        --model_preset=multimer                  \
        --db_preset=full_dbs                     \
        --output_dir=$PROJDIR/output             \
        --use_gpu_relax=True                     \
        -v 1
```

Under the hood, AlphaFold first carries out multiple sequence alignments (MSAs) of similar proteins to determine which amino acid residues of the protein are co-evolving and thus are probably next to each other in the 3D structure. AlphaFold next uses 3D structure templates. The results of iterations of these deep learning activities are a few different candidate 3D protein structures predictions, in PDB-formatted files (whose file extension is .pdb). You could look at and use these "unrelaxed" results, but they need refining. The final step that AlphaFold performs is a nonmachine learning optimization using the older scientific technique of "molecular dynamics" to produce the optimized "relaxed" PDB structure files, which are then ranked to produce the final "ranked" PDB structure files.

Viewing and Comparing AlphaFold Results

The output from AlphaFold is a 3D protein structure model encoded in a PDB file called ranked_0.pdb. There may be additional models, in output files ranked_1 .pdb, ranked_2.pdb, etc. Let's look at the structure in ranked_0.pdb.

A web page is now available that will allow you to view a PDB file in your browser without having to install any software. Go to www.ncbi.nlm.nih .gov/Structure/icn3d/full.html. Choose File, then Open File, then PDB File (appendable). Click Choose Files and choose the ranked_0.pdb output file from AlphaFold to be uploaded for viewing. Then click Append. The protein structure will appear displayed in the browser (Figure 11.1). Moving your cursor will rotate the structure, allowing you to view it from all angles.

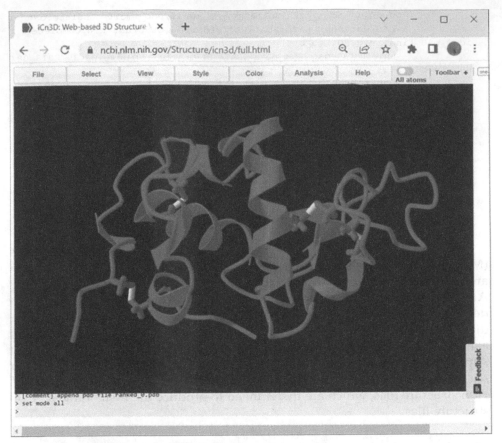

Figure 11.1: A view of an uploaded protein 3D structure model

If you want to look at the PDB structure on your local computer without going through the Internet, there are several PDB viewer applications you can download for free. One of them is Chimera at `www.cgl.ucsf.edu/chimera/download.html`. Installation is simple: download the installer and click it to run the install. Let's say that you have two protein structures you'd like to compare. One is the normal wild-type protein that you are probably able to download from PDB, and the other is the protein with a mutated amino acid residue whose model you may have obtained by running AlphaFold. In this example, we will look at the CFTR gene implicated in the cystic fibrosis disease. The wild type is PDB entry 2BBO. An AlphaFold wild-type model is also available. All PDB models already available for a given human protein are listed in the Structure section of the protein's Uniprot page at `www.uniprot.org`. The mutated protein in this exercise is the Phe508Del mutant. This mutation is a common cause of cystic fibrosis. In this particular case, a PDB model does exist: 1XMJ. However, if the mutated model doesn't already exist, you can make the change to the amino acid sequence in a FASTA file and run that FASTA file through AlphaFold to obtain the mutant 3D structure. Here is a comparison of a portion of the two FASTA sequences for CFTR. They are the same as each other (text in green) until residue Phe508, shown as "F" in red. This Phe508 residue is missing in the mutant sequence.

```
IKHSGRISFCSQFSWIMPGTIKENIIFGVSYDEYRYRSVIKACQLEEDISKFAEKDNIVLGEGGITLSEG
QQAKISLARAVYKDADLYLLDSPFIKHSGRISFCSQFSWIMPGTIKENIIGVSYDEYRYRSVIKACQL
EEDISKFAEKDNIVLGEGGITLSEGQQAKISLARAVYKDADLYLLDSPFG
```

To compare the two structures, open Chimera, choose File and then Open, and navigate to the first structure file 2bbo.pdb, which will then be displayed in beige on the Chimera screen. Open the second structure file by choosing File and then Open and navigating to and clicking 1xmj, which will display the structure in blue. Left-clicking the mouse and moving it will allow you to rotate and inspect the two structures that are side by side and not easy to compare. Line them up for comparison by choosing Tools ➪ Structure Comparison ➪ MatchMaker. Under Reference Structure, click 2bbo.pdb. Under Structure(s) To Match, click 1xmj.pdb. Then click OK and wait for the two structures to be aligned. They seem to overlap nicely for most of the protein. To see where Phe508 is, click Tools ➪ Sequence ➪ Sequence and then click 2bbo.pdb. Select the F at position 508, as shown in Figure 11.2, and then click Quit.

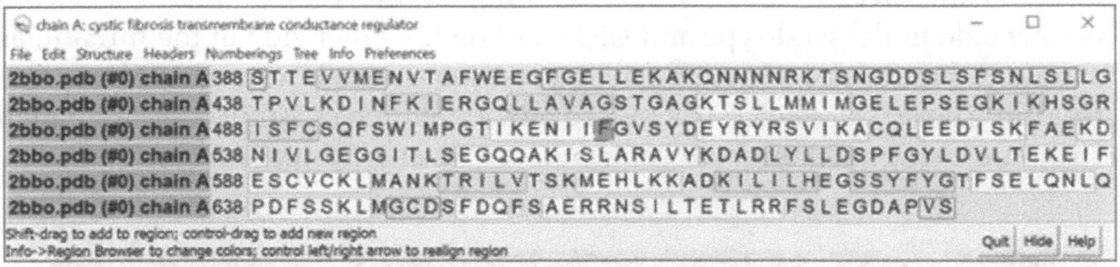

Figure 11.2: The sequence of PDB structure 2BBO with residue Phe508 selected and thus highlighted

Back on the structure display screen, a tiny portion of the beige structure will be highlighted with a bright green outline. To make that Phe508 residue more visible, select Actions ➪ Atom/Bonds ➪ Show, then select Actions ➪ Atom/Bonds ➪ Ball & Stick, finally again select Actions ➪ Color ➪ Forest Green. You can see in Figure 11.3 that the lighter wild-type and blue mutant structures diverge at the position of Phe508, because the mutant structure does not have Phe508.

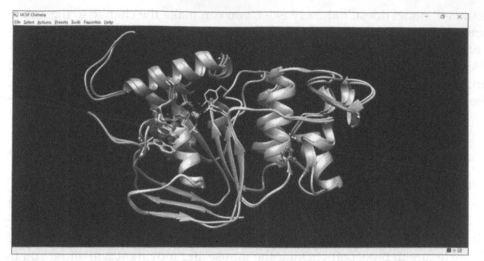

Figure 11.3: A comparison of the structures of wild-type (PDB 2BBO shown lighter) and Phe508Del mutant (PDB 1XMJ, shown darker) CFTR protein. Residue Phe508 is not present in the mutant structure. The mutant protein structure deviates from that of the wild-type at this position.

To continue comparing, again click Tools ⇨ Sequence ⇨ Sequence ⇨ 2bbo.pdb, and select more of the sequence around Phe508, selecting IFGVS and then Quit. Highlight those residues by selecting Actions ⇨ Atom/Bonds ⇨ Ball & Stick. Do the same for IGVS in 1xmj.pdb. You can then remove the bright green selection outline with Select ⇨ Clear Selection. You will see that residue Val510 sticks out on one side in the wild-type and sticks out on the other side in the mutant, as shown in Figure 11.4. Hovering over a residue shows information about that residue. The lack of Phe508 has caused a reorientation of Val510, and this small structure change is thought to contribute to causing the disease cystic fibrosis.

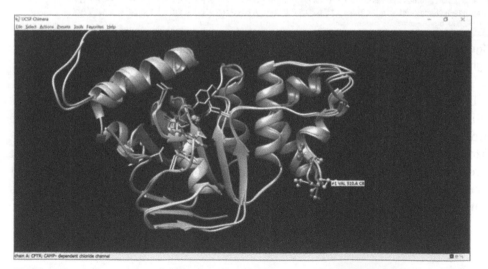

Figure 11.4: A comparison of the structures of wild-type (PDB 2BBO, lighter) and Phe508Del mutant (PDB 1XMJ, darker) CFTR protein. Residue Val510 has reoriented the mutant. Instead of facing left as in the wild type, it is facing right in the mutant.

This mutation causes a change in the surface, which can impact interactions with this protein. To view the surface, first remove what is currently displayed by selecting Select ⇨ Select All, and then Actions ⇨ Ribbon ⇨ Hide and Actions ⇨ Atoms/Bonds ⇨ Hide.

Let's first look at the wild-type surface with Select ⇨ Chain ⇨ A ⇨ 2bbo.pbs. Then select Actions ⇨ Surface ⇨ Show to produce Figure 11.5.

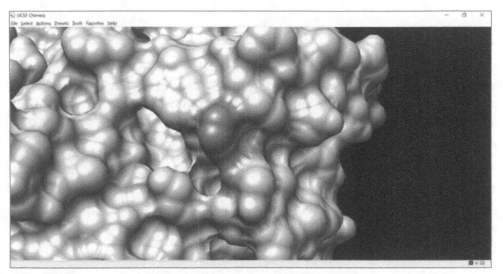

Figure 11.5: The surface of wild-type CFTR in the vicinity of Phe508

Hide the current display and repeat the previous process for displaying the surface of the mutant in 1xmj.pdb to produce Figure 11.6. We see that the surface topology is not the same as for the wild-type protein.

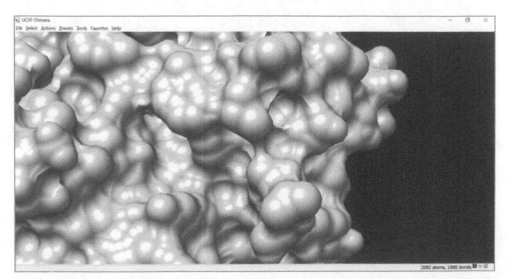

Figure 11.6: The surface of Phe508Del mutant CFTR in the vicinity of where Phe508 would be if it were not deleted

Such a change may adversely impact this protein's interaction with other protein partners, affecting function. The PDB viewer Chimera allows us to study the differences between two proteins, and AlphaFold enables us to generate 3D model predictions for any protein we are interested in. This allows the study of the differences in structure between healthy and mutant proteins, facilitating understanding of the disease mechanism and the design of drug treatments.

Summary

This chapter looked at applications of the analysis of genomes. We first discussed polygenic risk scores, which capture the additive effects of thousands of variants to influence our complex traits; it is rarely a single gene that causes a single trait. We then looked at metagenomics, which is used to analyze genetic material taken in bulk from material such as seawater, sewage, or soil, and explored how metagenomics can be applied to human samples. We also looked at how AlphaFold uses machine learning to predict new protein structures based on known information.

In the next chapter, we'll look more closely at some of the applications of genomics that are helping people to an enormous extent and have the potential to do even more good: those that have to do with cancers.

# CHAPTER 12

Cancer Genomics

In this chapter, we discuss the analysis of somatic mutations, which are mutations we develop across our life span. By far, the most studied somatic mutations are the mutations found in cancerous tumors, so we named this chapter "Cancer Genomics." However, the workflow presented in this chapter could also be applied to any tissue in the body, such as the analysis of your sun-damaged skin (provided you are willing to submit yourself to a punch skin biopsy) or the analysis of a benign polyp of your cat.

Somatic Genomes

As we age, our genome accumulates mutations and diverges from our birth genome. As mentioned in Chapter 2, our birth genome, derived from the fusion of the germ cells of our parents, is called our *germline genome*. Across our life span, our cells acquire mutations through replication errors or environmental factors such as UV rays, tobacco, toxins, or viruses. These acquired mutations are called *somatic mutations*, and by definition they affect only certain cells of our body.

This is the fundamental difference: germline mutations affect all our cells, whereas somatic mutations are present only in some cells of our body. Organs that accumulate more somatic mutations are those most exposed to environmental mutagens or with a high cellular turnover, like the skin, lung, gastrointestinal tract, liver, and blood. In contrast, cells of the brain and muscles, for example, are less prone to mutations.

Somatic mutations can also develop while in utero. So while our germline genome is most often referred to as our *birth genome*, it would be more accurate to call it our genome at fertilization. In practice, a mutation that develops in the first cell divisions of the fertilized egg will affect the majority of our cells and is a nearly constitutional or germline mutation. A recent study from Iceland showed that identical twins differ on average by 5.2 such near-constitutional mutations, which are very hard to distinguish from germline mutations (Jonsson et al., 2021). Somatic mutations that occur later in embryonal development can create mosaicism, where an individual is made of two or more populations of cells with different genomes; an example can be someone born with a large dark birthmark such as a congenital melanocytic nevus, typically caused by a mutation in the NRAS or BRAF mutation during skin development.

Cancer

As our cells accumulate somatic mutations through replication errors or environmental factors, some somatic mutations can confer a proliferation or fitness advantage to a cell over our other cells. This mutated cell will divide and multiply into identical cells with the same mutated genome (that is the same somatic genome), forming a *clone*—a population of cells with the same genome. This can cause, for example, the development of benign skin growths as we age. A benign growth can turn malignant or cancerous if it acquires the capability to metastasize and invade other tissues. Cells of the original clone can then mutate to form new clones with new characteristics.

Let's go deeper in what characterizes cancer cells. Historically, cancer was described as a disease caused by oncogenes and mutations in tumor suppressor genes.

Oncogenes

An *oncogene* is a gene that has the potential to cause cancer if it is mutated or overexpressed. Oncogenes are typically genes fostering cell growth. Events that make them turn "bad" include the following:

- **An activating mutation:** If a gene codes for a cell receptor, an activating mutation can render the receptor constitutionally activated, even if the molecule it is supposed to bind (called a *ligand*) is not present. If the mutated gene product is a protein kinase, which is a protein that acts like an "on" or "off" switch by attaching a phosphate group, then the protein kinase might be permanently switched on or off. Kinases can be cell receptors or operate in the cytoplasm. As an example, activating mutations of the epidermal growth factor receptor (EGFR) gene, which codes for a cell receptor that has a tyrosine kinase domain, can drive many cancer types. Gain of function mutations can enhance the activity of the resulting protein, such as increasing its enzymatic activity.

- **Gene fusions:** Gene fusions caused by chromosomal translocations can result in a hybrid protein that is constitutively active. The best known gene fusion is caused by the fusion of the BCR gene (from broken chromosome 22) with the ABL1 gene (from broken chromosome 9); this genetic abnormality, known as the Philadelphia chromosome or translocation, causes the hybrid protein kinase BCR-ABL1. This hybrid protein is always switched on, which drives the development of chronic myelogenous leukemia and other leukemias.

- **A gene amplification:** Some genes can have their DNA sequence copied several times in the genome of cancer cells. This abnormality can result, for example, from errors in DNA replication or DNA repair or movements of mobile genetic elements such as transposons. As a result, cells can have up to several hundred extra copies of a gene, which is called *gene amplification*. It is a type of copy number variation (CNV). The extra copies can be adjacent to the normal gene or in a different chromosome or even in extra chromosomal circular DNA structures. These extra copies dysregulate the normal operating mode of the genome and can help the transformation of normal cells into cancerous cells.

    About 30 percent of cancers have an amplification of at least one member of the multifunctional MYC transcription factor family (MYC, MYCL, or MYCN).

- **Dysregulation:** Mutations in the promoter or enhancers of the gene, translocations that move the gene just downstream of a very active promoter, and epigenetic changes as well as changes in the regulatory network of the gene can lead to overexpression of the gene product, or expression at inappropriate times. Gene amplification, as we saw previously, can also cause dysregulation. Mutations that affect the stability of the resulting mRNA or protein can lead to excess mRNA or protein product.

Tumor Suppressors

A *tumor suppressor* is, as you can guess, a gene that inhibits tumor development. Tumor suppressor genes include the following:

- Genes that act as a brake on cell proliferation, such as the retinoblastoma (Rb) gene and the cyclin genes that regulate and block the progression of cells in the cell cycle, or the adenomatous polyposis coli (APC) gene that controls cell proliferation and is mutated in 35 to 70 percent of colon cancers.

▪ Genes involved in DNA repair, like the well-known BRCA1 and BRCA2 genes that control the repair of DNA double-strand breaks and are associated with breast cancer. More than 200 genes are involved in DNA repair and perform functions such as base excision repair, mismatch excision repair, nucleotide excision repair, nonhomologous end joining, or direct DNA repair.

▪ Genes regulating cell death. Normal cells must undergo programmed cell death (apoptosis) or be forced into senescence (a state in which they cannot divide anymore) whenever they accumulate mutations or abnormalities that could put the whole organism at risk. This tight control is key in all multicellular organisms: each cell must be a team player ready to die on demand. Cancer cells have the key capability to evade cell death signals, and so can grow uncontrollably, even if it leads to the death of the host. Cancer is sometimes described as a reversal from multi-cellularity to mono-cellularity, where it is "everyone for themselves." The most important pro-death gene is the transcription factor p53. The p53 protein (the number 53 refers to its mass in kilodaltons, the dalton being a very small unit of mass, equal to that of a carbon-12 atom under certain conditions) is also called the *guardian of the genome*. If a cell faces abnormal conditions or is under stress, p53 can temporarily stop the cell from dividing, induce changes in cell metabolism, or trigger senescence. If the cell damage is irreparable, p53 will trigger programmed cell death. p53 is mutated in more than 50 percent of sequenced cancers, with a higher prevalence in advanced cancers, and is a common driver of cancer relapses.

The Promise and Reality of Cancer Precision Medicine

Although it is easy to kill a cancer cell—just treat it with bleach or any other toxic compound—the difficulty lies in killing cancer cells while sparing normal cells. Chemotherapy drugs work by killing all dividing cells, whether cancer fast-dividing cells or normal dividing cells; they are therefore not selective and in consequence cause a lot of toxicity for the patients. The concept of cancer precision medicine is to devise drugs that target genetic mutations or mechanisms that are specific to cancer cells and hence spare healthy cells. The idea is to sequence the genome of the tumor of each new cancer patient, identify key mutations, and prescribe drugs exquisitely targeting cells with these mutations while sparing normal cells. There are two main classes of targeted drugs: small molecules and monoclonal antibodies. The latter class consists of synthetic antibodies that target antigens (proteins recognizable by the immune system) present on the surface of cancer cells.

The first small-molecule cancer-targeted drug, imatinib, approved and sold under the brand name Gleevec since 2001, targets the hybrid BCR-ABL protein that causes most cases of chronic myelogenous leukemia. This fusion protein, caused by chromosomal translocation, is not present in normal cells, so cancer cells are selectively targeted. Imatinib became a huge success: while CML is a deadly disease, imatinib brings remission to 98 percent of patients. Patients after two years on remission with imatinib have the same life expectancy as people without cancer. Around the same time the first monoclonal antibody drugs, such as trastuzumab (Herceptin), were approved. Trastuzumab targets the protein human epidermal growth factor receptor 2 (HER2), which is very present at the surface of breast and gastric cancer cells.

Scientists began working on replicating this success story to other cancer types and struggled. Key factors explain the difficulty of replicating this success.

- Very few cancers are caused by a single genetic abnormality like in the case of CML. Most cancers are caused by several mutations that can be specific to each patient. So this approach requires pharmaceutical companies to develop a large number of drugs able to target the key cancer driver mutations of each patient.

- Cancer cells mutate fast and create new clones. The same primary tumor of a patient will contain several clones. Most tumor biopsies capture only one or two clones. Multiple biopsies throughout the tumor are more likely to capture several clones but are not always feasible for the patient. So the sequenced tumor genome might miss the key mutations of an aggressive clone that was not sampled during the biopsy.

- Even if all clones of the primary tumor are captured in the biopsy, these clones will evolve through time. In particular, treatment creates a selective pressure that pushes clones to evolve ways to resist the treatment and create new treatment-resistant clones with different sets of mutations.

- If the primary tumor spreads to new locations, each metastasis can develop new clones with completely different mutation profiles that were not present in the original tumor.

- As cancer cells defy all rules and evade programmed cell death (the rule is "there are no rules; everyone for themselves"), several drugs are often required to target a single clone. Combinations of drugs can aim to target several pathways or create a double hit on the same pathway. If there are metastases, this leads to a large number of drugs and a great risk of toxicity for the patient. Some cancer centers are more adventurous than others and prescribe a large number of combination drugs at reduced dosage, taking the risk of unknown drug combination toxicity. Other centers limit treatments to tested drug combinations but will not target some driver mutations that might enable the cancer cells to evade treatment.

■ There is no guarantee that detected cancer-driving mutations matter at all. A pathway might be broken because of mutations upstream of the detected mutation. The cancer cell can then rewire itself to use alternate pathways and be oblivious to the drug targeting the already broken pathway. We need to keep in mind that only half of human genes are assigned a molecular or biological function, and less than 15 percent of human genes have a known pathway. If we already struggle to understand the behavior of a normal cell, it will be harder to predict the behavior of a cancer cell with a large number of mutated genes and duplicated or missing entire chromosomal regions.

■ The bioinformatics analysis of tumors is complex. Cancer cells are mixed with normal cells, so the variant callers need to take into account an assumed percentage of normal cells to properly call the tumor mutations. In the case of stem cell or bone marrow transplants, cancer cells are mixed with the normal cells of the patient and the cell of the transplant donor and need to be disentangled. Some genes might be aberrantly over- or under-expressed not because of mutations but consecutively to abnormal epigenetics marks. RNA analysis is recommended to identify which pathways are operational in a heavily mutated cell. So a comprehensive analysis should include the DNA, RNA, and epigenetic analysis of the main tumor and of each of the metastases and be repeated regularly throughout the treatment as the cancer clones mutate and evolve.

Some patients respond remarkably well to targeted drugs and achieve durable partial or complete remission. Many patients, however, either do not have mutations for which drugs exist or do not improve with drugs they "should" respond to on the basis of their reported tumor mutation profile.

In front of all this complexity, scientists looked at a different approach and in particular at how we can harness our immune system to eliminate cancer cells. This approach, called *immunotherapy*, is not new. The immune protein interferon type I was investigated since the 1970s against cancer but was abandoned because of limited efficacy and severe side effects. The immune-signaling protein interleukin 2 is administered in certain cancers like neuroblastoma. The tuberculosis BCG vaccine is injected in the cancerous bladder of patients to trigger an immune reaction to help clear the cancer and is now part of the gold standard treatment for early-stage bladder cancer. The last 10 years saw two major breakthroughs in immunotherapy: CAR T cells and immune checkpoint inhibitors.

CAR T cells are cells extracted from patients, reengineered to attach to an antigen protein on the surface of the tumor, expanded in vitro, and reinjected to the patient. After reinjection, these cells zoom to the tumor to kill cancer cells. This technique, while promising, is still in early stages, particularly for the management of serious side effects after reinjection. It has been approved by the FDA for acute lymphoblastic leukemia (ALL) in children and certain lymphomas in adults. It is currently unknown if this technique is applicable to solid cancers, such as breast or prostate cancers.

Immune checkpoint inhibitors represent an exciting breakthrough that disables a ruse used by cancer cells to evade the immune system. To avoid attacking our own body and developing autoimmune disease, our immune system must respect immune checkpoints. T cells have receptors on their outer surface called PD1 and CTLA4. If any of these receptors is activated by its partner protein (PD-L1 for PD1, CD80/CD86 for CTLA4), an "off signal" is sent to the T cell, which dampens the immune response. Cancer cells express the ligands of PD1 and CTLA4 to act as part the body, effectively disguising themselves and avoiding being destroyed by the immune system. Immune checkpoint inhibitors, which inhibit either PD1, PD-L1, or CTLA4, revolutionized the treatment of metastasized melanoma and are being progressively extended to other tumor types such as lung cancer (NSCLC).

Immune checkpoint inhibitors work best when these two conditions are met:

- The tumor is very mutated, as quantified by the tumor mutation burden (TMB), which is the number of somatic mutations per million bases. A mutated tumor is more likely to appear foreign and be attacked by the immune system, so the more mutations, the better. This explains the success of immune checkpoint inhibitors in melanomas (always very mutated by UVs) and tumors with genomic instability (such as microsatellite instabilities, which we will evaluate later in this chapter).

- The tumor is infiltrated by immune cells. Immune checkpoint inhibitors make it more difficult for tumors to evade immune cells but are pointless if there are no immune cells in the tumor in the first place.

A third criterion, expression by the tumor of the immune checkpoint ligands, is not always reliable, as the production of PD-L1 protein by a tumor can vary according to space and time.

Somatic or Germline? Cancer Predisposition

We talked previously about somatic mutations in oncogenes and tumor suppressors. Some individuals can have these mutations in their germline. As an unfortunate consequence, these individuals will be predisposed to benign and cancerous growths.

The most deleterious mutations are rare: germline p53 mutations are present in between 1 in 5,000 to 1 in 20,000 individuals and cause Li-Fraumeni syndrome. Affected families have a high risk of cancer in childhood and early adulthood and require close follow-up. Individuals older than 40 are not usually tested as germline p53 mutations generally cause cancers before that age. Germline p53 mutations typically cause sarcomas, brain cancers, breast cancers, and acute leukemia, and they are suspected in presence of rare subtypes of these cancers. Affected individuals are highly susceptible to radiation. Several programs are available worldwide to provide yearly whole-body MRIs with radiologists trained to detect typical Li-Fraumeni tumors and to avoid false positives.

Besides p53 mutations, most mutations predisposing to cancer are mutations affecting DNA repair genes. The most common hereditary cancer disposition is the Lynch syndrome, caused by a germline mutation in any of the DNA mismatch repair genes (such as MSH2 to MSH6, or PMS2). People with the Lynch syndrome have a high risk of colorectal cancer before age 50, as well as a higher risk for several other cancers.

Pathogenic germline mutations in the BRCA1 or BRCA2 are well known to increase the risk of breast and ovarian cancer, in particular if they are associated with a family history of such cancers. The lifetime risk of developing breast cancer for women in the general population is about 13 percent; the breast cancer risk increases to 55–72 percent with a pathogenic BRCA1 mutation, and to 45–69 percent with a pathogenic BRCA2 mutation, by age 70–80 years old. For ovarian cancer, the risk increases from 1.2 percent (general population lifetime risk) to 39–44 percent (pathogenic BRCA1 variant) and 11–17 percent (pathogenic BRCA2 variant) by age 70–80 years old. The BRCA gene status (germline and somatic) is frequently tested in patients with breast or ovarian cancer, as it affects both prognosis and choice of treatment.

Overall, however, and although a lot of research has been devoted to genetic cancer predisposition, only 15 percent of cancers are considered hereditary. The vast majority of cancers (85 percent) have environmental causes.

Chromothripsis

Chromothripsis comes from the Greek *thripsis*, which means "shattering." It was first discovered in 2011 and refers to a single catastrophic event that shatters a chromosomal area into many pieces. DNA repair mechanisms attempt to fix the shattered chromosome, but the repairs are imperfect. As a result of this single event, the chromosome can display duplications, deletions, inversions, translocations, and copy number variations. Chromothripsis is rare in the germline and causes congenital disease. In contrast, it occurs in about 40 percent of cancers, with the highest prevalence in sarcomas (70–100 percent), melanomas, glioblastomas (brain tumor of glial cells), and lung adenocarcinomas (about 50 percent for each).

Detecting chromothripsis from WGS requires detecting all SVs and CNVs and applying a criterion incorporating the number of overlapping SVs, the pattern of alternating CNV (usually only two, or sometimes three, CNV states), and DNA breaks.

Epigenetics of Cancer

We saw that the genome of cancers can develop large abnormalities, such as amplification of genes, translocations that match genes with the wrong promoter, or creation of fusion proteins that are abnormally activated. How does it happen? Cancer cells are genetically unstable. One of the key contributors for this instability is the *hypo-methylation* of the genome of cancer cells. More than 50 percent of the human genome is composed of mobile elements, and the proportion of mobile elements in most higher organisms' genomes is similarly high. These mobile elements are mainly transposons (often described as *jumping genes*) and endogenous retroviruses that entered the genome of our ancestors up to millions of years ago. To prevent these mobile elements from changing position in the genome and creating mutations, the mobile elements are silenced by DNA methylation: a chemical methyl group gets attached to the C (cytosine) base of the mobile elements, particularly in areas with consecutive CGs (called *CpG islands*). In cancer, the methyl groups disappear, so the transposons and endogenous retroviruses are not silenced anymore; they can move or duplicate to other areas of the genome and cause mutations. Hypo-methylation in the promoters or enhancers of oncogenes can also aberrantly increase the expression of oncogenes.

In addition, the promoters and enhancers of tumor suppressors are often hyper-methylated, which silences them. The important PTEN tumor suppressor, which counteracts several oncogenes in proliferation pathways, is not expressed in up to 70 percent of prostate cancer patients; this loss is sometimes due to PTEN mutations but most often caused by aberrant methylation of the PTEN promoter.

Histone marks (acetyl and methyl groups placed on the tail of the histone proteins around which DNA is wrapped) are also deregulated in cancer and contribute to aberrant gene expression. Repressive histone marks can aberrantly remain on developmental genes and keep cancer cells in an undifferentiated (stem-cell- or progenitor-cell-like) and highly proliferative state.

So, in summary, the genome of cancer cells is globally hypo-methylated, which de-represses the mobile elements and increases genomic instability. Tumor suppressors are aberrantly silenced; oncogenes can be de-methylated at inappropriate times; and histone marks, which provide an additional way to control transcription, are deregulated.

These epigenetic abnormalities, which help drive cancer, cannot be picked up by WGS or WXS, as they do not change the DNA letters. They can be evidenced only by the study of the following:

▪ **RNA expression:** The extracted RNA can be sequenced via RNA-Seq in an approach similar to WGS, or to reduce costs, hybridized to microarrays. RNA-seq is a more powerful technique, as it allows not only detecting abnormalities in gene expression but also detecting gene fusions and studying isoforms and patterns of alternate splicing.

- **DNA methylation:** If DNA is treated with the chemical bisulfite, the methylated cytosines are left untouched while the unmethylated cytosines are converted into uracil. The sequencing of the resulting DNA reveals the position of all methylated cytosines throughout the genome. This technique is called *whole-genome bisulfite sequencing* (WGBS). A cheaper and more popular alternative, particularly in the clinic, is to hybridize the bisulfite-treated DNA to a microarray; the most popular methylation array is the MethylationEPIC array from Illumina, which allows the quantitation of methylation at about 850,000 CpG sites.

- **Histone marks:** These can be studied by the technique chromatin immunoprecipitation-sequencing (ChIP-seq). Cells are treated with formaldehyde to cross-link their DNA with all proteins bound to it (such as histones and transcription factors). The histone marks are identified ("immunoprecipitated") by adding antibodies specific to the histone marks of interest. The bound DNA is sequenced and corresponds to the areas where the histone marks of interest are located on the genome. If antibodies specific to transcription factors are used instead, this technique allows determining where transcription factors bind in the genome.

Note that the Nanopore sequencing platform can allow users to perform WGS and study DNA methylation in one step, as methylated cytosines trigger a different signal as they travel through a nanopore. The drawbacks are a high error rate in WGS and a higher rate of false positive (false call of methylated C when it is unmethylated) versus WGS.

Mechanisms of Cancer

Cancer used to be seen as a genetic disease of bad genes (oncogenes) against good genes (tumor suppressors). It has become progressively clear that it is much more complicated. It is difficult from the genome alone to differentiate a cancerous tumor from a benign tumor, or even from aged tissue. A recent study analyzed the excess skin removed from individuals who underwent eyelid cosmetic surgery. Scientists were surprised to see that the sun-aged skin of these individuals had oncogene mutations that most thought were specific to cancers (Martincorena et al., 2015). Robert Weinberg and Douglas Hanahan published in 2000, and revised in 2011 and 2022, a landmark article on the hallmarks of cancer. In this article, Weinberg went beyond the oncogene/tumor suppressor dichotomy and summarized the cellular and biological mechanisms behind cancer. Although by nature imperfect and always a work in progress, it is always an excellent basis for explaining cancer mechanisms.

Figure 12.1 shows the main hallmarks of cancer, as defined by the 2022 revision of the landmark publication of Weinberg and Hanahan. Note that of the 14 hallmarks, 13 could also apply to benign tumors, and only one is specific to cancer: invasion and metastasis.

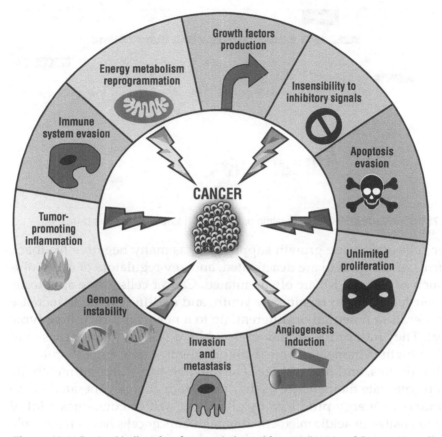

Figure 12.1: Revised hallmarks of cancer (adapted from Hallmarks of Cancer: New dimensions, Hanahan, 2022)

Cell proliferation is tightly regulated in our bodies. Sudden proliferation of skin cells, for example, should exclusively occur to repair a wound and so requires growth factors secreted for that purpose. Secreted growth factors activate receptors located on the surface of the cell, which trigger a cascade of chemical reactions between proteins called a *signaling pathway*. This eventually makes a cell re-enter the cell cycle and multiply. Cancer cells can evade this strict dependence on growth factors by either secreting themselves the growth factors (or enrolling other cells to secrete them) or developing mutations to have either the growth factor receptors or the signaling pathways permanently activated.

It is a good idea to be familiar with some key signaling pathways controlling cell proliferation, such as the PI3K-AKT-mTOR pathway, as it is an important driver of cancer and is targeted by many cancer drugs, as illustrated in Figure 12.2.

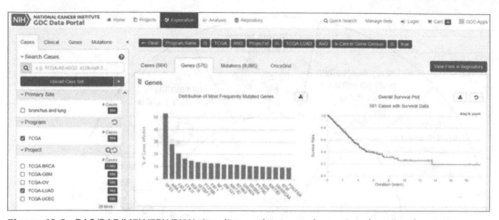

Figure 12.2: RAS/RAF/MEK/ERK/PI3K signaling pathways and associated cancer drugs

Cancer cells also evade growth suppressors, as many negative feedback mechanisms of cell growth are deactivated, and key regulators of cell proliferation such as p53 and RB are often mutated. Cancer cells evade apoptosis, extend their telomeres to regain new youth, and dedifferentiate to increase their proliferation potential or differentiate to a new cell type (phenotypic plasticity). They induce the growth of new blood vessels (angiogenesis) to feed their growth, remodel their microenvironment, secrete metalloproteases to dissolve tissues, and metastasize. They switch to a different metabolic program to generate their energy, as they shift from energy generated in the mitochondria to energy produced by glycolysis, which consumes a lot of glucose and creates an acidic microenvironment. Cancer cells have an unstable genome that mutates fast: DNA repair pathways are damaged, and large areas of the genome that are usually silenced by methylation get suddenly active and trigger genome instability and mutations. The chromatin of cancer cells gets remodeled, which changes the regulation of gene expression in cancer cells even in the absence of mutations. Cancer cells that stop replicating and turn senescent secrete inflammatory factors that can stimulate cancer tumor development. Resident bacteria, fungi, or viruses in the tumor microenvironment can influence the growth of cancer cells and constitute an emerging area of cancer research.

Importantly, cancer cells evade the immune system. While they maneuver to be seen by the body as an inflamed wound that needs to be fed with growth factors and nutrients, cancer cells create an immunosuppressive environment and evade the cytotoxic T lymphocytes and NK cells that are supposed to kill them.

Samples

The analysis of a somatic genome, such as a tumor genome, requires two samples.

- The tumor sample, such as a cancer biopsy
- A normal sample corresponding to the germline genome, which the tumor sample will be compared to

The bioinformatics workflow will identify somatic mutations, here tumor mutations, by searching for mutations that are present in the tumor biopsy but not in the germline reference sample. The reference sample is usually a peripheral blood sample or a sample of normal tissue adjacent to the tumor.

In the rare cases where it is impossible to obtain a reference normal sample, the tumor sample can be compared to a data set containing the most common variants in a population of similar ethnicity; this approach, however, generates a lot of false positives for the detected tumor mutations and is not recommended.

The cancer sample is ideally a tissue immediately frozen after biopsy so that the DNA is of best quality. In practice, tumor samples are often preserved with formalin and embedded in a paraffin wax block, a technique called *formalin-fixed paraffin-embedded* (FFPE). Formalin, which is another name for formaldehyde, fragments and damages the DNA and can deaminate cytosine into uracil or thymine; this results, after sequencing, in some Cs changed to Ts (or on opposite strand: G changed to A). Deamination is more common in samples with high levels of DNA fragmentation. The sequencing laboratories evaluate the DNA quality prior to sequencing and usually report the level of DNA fragmentation.

Tumors are very heterogeneous and often harbor different clones. To be able to capture the genetic diversity of each clone, it is important to pick cells in different regions of the tumor.

Somatic Variant Analysis

Now that we understand the key mechanisms of cancer, let's analyze a cancer genome. As stated previously, we need two genomes.

- The *germline* genome of the individual to be studied, typically sequenced from a blood sample
- The genome of the cancer of this same individual, also called *somatic* genome

The germline cancer will allow us to distinguish the mutations the individual was born with from the new mutations developed by the cancer. In addition, the germline genome will enable the automated correction of sequencing artifacts. The comparison of the two genomes (germline and somatic) and consecutive determination of the cancer mutations is done by software called a *somatic variant caller*.

Several somatic variant callers are available, such as Strelka2, Mutect2, VarScan2, or MuSE. Strelka2 and Mutect2 tend to be the most popular. Recently published benchmarks show that Strelka2 and Mutect2 perform similarly, with Strelka2 slightly more accurate in more mutated tumors (more than 20 percent mutations) and Mutect2 marginally better in tumors with less than 10 percent mutations. As Strelka2 runs on average 20 times faster than Mutect2, we will use Strelka2.

Somatic variant callers not only compare the somatic versus germline genomes but also allow for variable *tumor purity*. Tumor purity is defined as the percentage of tumor cells in the tumor samples, as biopsied tumor samples contain a mix of cancerous and normal cells. Strelka2 can be used even if the tumor purity is only 5–10 percent and allows for up to 10 percent of contamination of the normal sample with tumor cells.

Strelka2 uses as input BAM files or its compressed alternative CRAM files. This means it works from genomes that have already been aligned to a reference genome. The steps to produce these BAM files are exactly what we covered in previous chapters.

1. Quality control of the FASTQ file

2. Alignment to a reference genome with BWA

3. Marking of duplicate reads

Let's suppose we have accomplished these steps for the germline and cancer FASTQ files and produced a germline file `normal.bam` and a somatic file `tumor.bam`.

Strelka2 can detect SNVs and indels of a maximum size of 49 bases or less. To make the calling of variants more accurate, Strelka2 is designed to run after an initial step performed by a software package called Manta that calls structural variants and large indels. Both Strelka and Manta have been designed by the company Illumina. At the end of the analysis, we will have called all tumor SNVs, small and large indels, and structural variants.

Let's install Manta and Strelka. Go to `https://github.com/Illumina/strelka/releases` and `github.com/Illumina/manta/releases`, save the links of the latest binary releases of Strelka and Manta (the files work for most Linux distributions), download the files with the command `wget` (paste the links after `wget`), and then unpack the files with `tar` and install package additions.

```
> wget https://github.com/Illumina/strelka/releases/download/v2.9.10/
  strelka-2.9.10.centos6_x86_64.tar.bz2
> wget https://github.com/Illumina/manta/releases/download/v1.6.0/
  manta-1.6.0.centos6_x86_64.tar.bz2
> tar -jxf strelka-2.9.10.centos6_x86_64.tar.bz2
> tar -jxf manta-1.6.0.centos6_x86_64.tar.bz2
> apt-get update -qq
> apt-get install -qq bzip2 gcc g++ make python zlib1g-dev
```

Manta and Strelka2 are each executed by first running a configuration script and then an execution script. These scripts are written in Python and are provided as part of the installation. The configuration scripts can be modified to change the configuration parameters. The default parameters are quite optimal, so we will not change the configuration scripts and run them as is.

Manta requires four parameters: a germline BAM file, a somatic BAM file, the genome reference that was used to align both BAM files, and the name of the directory where the analysis will be conducted. Manta compares the somatic (tumor) BAM to the germline BAM to detect what new structural variants and indels have emerged in the tumor. Like in Chapter 8, we use as a reference genome the GRCh38 version without alternate regions.

```
wget ftp://ftp.ncbi.nlm.nih.gov/genomes/all/GCA/000/001/405/
GCA_000001405.15_GRCh38/seqs_for_alignment_pipelines.ucsc_ids/
GCA_000001405.15_GRCh38_no_alt_analysis_set.fna.gz


ref_fasta=GCA_000001405.15_GRCh38_no_alt_analysis_set.fna

MANTA_INSTALL_PATH=your_code_directory/manta-1.6.0.centos6_x86_64/bin


#Run Manta configuration step
${MANTA_INSTALL_PATH}/configManta.py
--normalBam normal.bam \   #germline BAM file
--tumorBam tumor.bam \     #somatic BAM file
--referenceFasta $ref_fasta \#genome reference used to align both BAM files
--runDir ${MANTA_ANALYSIS_PATH}#directory where the anaylsis is conducted
```

Manta is now ready for execution. Manta will search, analyze, and score all tumor structural variants, medium-size indels, and large insertions. The execution is launched from the analysis directory specified in the previous step. We are using an AWS instance with eight cores and indicate this information as a parameter so that Manta parallelizes the analysis.

```
#List all Manta execution options (option -h)
${MANTA_ANALYSIS_PATH}/runWorkflow.py -h

#Execute Manta on an AWS instance with 8 cores
${MANTA_ANALYSIS_PATH}/runWorkflow.py -m local -j 8
```

Manta produces the following files as output in the directory ${MANTA_ANALYSIS_PATH}/results/variant:

- somaticSV.vcf.gz: List of scored structural variants and large indels.

- candidateSV.vcf.gz: List of unscored SVs and large indels with less stringent criteria.

- candidateSmallIndels.vcf.gz: List of candidates indels less than 50 bp. This file will be provided as an input for Strelka during our next step.

We are now ready to call the tumor SNVs and small indels with Strelka. First, we run the Strelka configuration step. In addition to the name of the germline BAM, somatic BAM, genome reference, and analysis directory, we indicate as input the `candidateSmallIndels` file produced in the previous step.

```
STRELKA_INSTALL_PATH=your_directory/strelka-2.9.10.centos6_x86_64/bin

#Run Srelka configuration step
${STRELKA_INSTALL_PATH}/bin/configureStrelkaSomaticWorkflow.py \
--normalBam normal.bam \
--tumorBam tumor.bam \
--referenceFasta $ref_fasta \
--indelCandidates ${MANTA_ANALYSIS_PATH}/results/variants/
candidateSmallIndels.vcf.gz \
--runDir ${STRELKA_ANALYSIS_PATH}
```

We can now run Strelka to call all SNVs and small indels. By default, Strelka scans the entire genome. Note that it is possible to restrict the search to specific regions of the genome by providing a BED file with the `--callRegions` option. In the following code, we run Strelka on the entire genome, so without the `--callRegions` option:

```
#List all Strelka execution options (option -h)
${STRELKA_ANALYSIS_PATH}/runWorkflow.py -h

#Execute Strelka on an AWS instance with 8 cores
${STRELKA_ANALYSIS_PATH}/runWorkflow.py -m local -j 8
```

Strelka produces the following files as an output in the directory `${STRELKA_ANALYSIS_PATH}/results/variant`:

- `somatic.snvs.vcf.gz`: All detected tumor SNVs
- `somatic.indels.vcf.gz`: All detected tumor indels

After annotation, the resulting VCF files list a large number of mutated genes. It is important to distinguish between driver and passenger mutations. *Driver mutations* confer a strong selective advantage to the cell and are the most important. Most of the mutations are *passenger mutations* and do not confer a significant proliferative advantage to the cell. Very long genes—for example, the TTN gene, which codes for the Titin protein, or the RYR2 gene, with 135 exons—frequently feature as mutated genes for the simple reason that a long gene is more likely to carry a mutation than a short gene, but they are not thought to have any effect on the growth of the cancer.

Many cancer centers focus on the genes of the Cosmic Cancer Gene Census produced by Sanger Institute. This list of cancer driver genes is available at `cancer.sanger.ac.uk/census` and is organized into tier 1 genes (genes most likely to drive cancer) and tier 2 genes (potential cancer driver genes with less evidence). The Cancer Census Genes are also classified according to the hallmarks of cancer we discussed previously. This list is evolving, so feel free to add your own candidate cancer driver genes.

A useful resource for then checking if the found mutations are usual or unusual for the cancer type is the Cancer Genome Atlas, better known as TCGA. Go to `portal.gdc.cancer.gov` and click the Exploration button.

The cancer types are abbreviated in TCGA with codes, such as LUAD for lung adenocarcinoma or COAD for colon adenocarcinoma. The full list is available at `gdc.cancer.gov/resources-tcga-users/tcga-code-tables/tcga-study-abbreviations`. Let's look, for example, at the most common mutations of the Cancer Census Genes in Lung adenocarcinoma. On the left side, under Program, click TCGA, and under Project, click TCGA-LUAD. We can see that there are currently 564 cases that have been sequenced. On the right side, select the Genes tab. You can now see, as in Figure 12.3, the list of most mutated genes by frequency in all 564 samples.

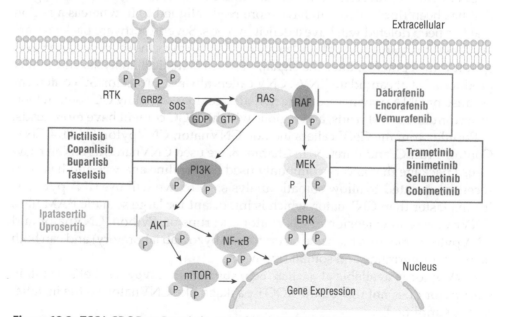

Figure 12.3: TCGA GDC Data Portal, the most frequently mutated Cancer Census Genes in Lung adenocarcinoma

Note that we can further refine the graph by filtering the type of mutations by selecting the Mutations tab on the left side. This Cancer Atlas is invaluable for querying cancer genomes and understanding mechanisms.

Up-to-date treatment protocols, including mutation targeted and immunotherapy protocols, are available at www.cancer.gov. Click Cancer Type, and then select your cancer of interest. The Patient version provides a good introduction; then go to the Health Professional version to list the treatment protocols and the latest trial results.

The list of FDA-approved drugs targeting Cancer Census Genes keeps evolving. A good list, including small molecules and monoclonal antibodies, is available at `www.mycancergenome.org/content/page/overview-of-targeted-therapies-for-cancer`. Not all drugs might be available in your local geography. They might, however, be available in clinical trials (`www.clinicaltrials.gov`).

Copy Number Changes

Cancer genomes can develop large alterations whereby large regions of the genome can be either duplicated once or more or deleted. These changes are referred to as *copy number variations* (CNVs). The changes can be associated with better or worse survival and can inform treatment for some cancers.

CNVs can be inferred from the variation in depth in a BAM file. A region that has been duplicated will have more reads aligned to it, whereas a region that has been deleted will have a deficit in reads. Several software packages are available that scan a BAM file, calculate the depth across the sequenced genome, split it into segments, filter the signal, perform statistical analysis, and finally produce a list of candidate CNVs. CNV callers always correct for GC content, as because of DNA polymerase characteristics, regions with low GC content tend to accumulate ewerf reads, and regions with high GC content have more reads.

Popular somatic CNV callers include CNVnator, CNVpytor, GATK4 CNV, Control-FREEC, and PureCN. In Chapter 8, we used CNVnator, as it is an easy-to-use package that is very commonly used in germline analysis and has been recently updated to allow somatic analysis. Here, we will use CNVpytor, as it runs faster than CNVnator, which is important for large somatic BAM files. CNVpytor is an extension of CNVnator and runs on Python. CNVnator and CNVpytor come from the same laboratory (Abyzovlab laboratory) and can both be used for germline and somatic analyses.

CNVpytor is available at `github.com/abyzovlab/CNVpytor`. Let's install it. CNVpytor does not rely on the ROOT package like CNVnator, so the installation is simplified.

```
# Installation CNVpytor
git clone https://github.com/abyzovlab/CNVpytor.git  #copy the git repository
cd CNVpytor
python setup.py install    #install
cnvpytor -version          #check CNVpytor installed correctly
```

We are now ready to use CNVpytor. The first step takes the longest: it extracts all necessary information from the tumor BAM and stores it into a file called here myfile.pytor. All subsequent commands will add further information to this myfile.pytor file. CNVpytor uses as input the tumor BAM file to calculate read depth. The names of the chromosomes must correspond to the names in the BAM file. CNVpytor automatically detects the reference genome.

```
# Extract information from BAM file
./cnvpytor -root myfile.pytor -tree tumor.bam -chrom chr1 chr2 chr3 chr4
chr5 chr6 chr7 chr8\
chr9 chr10 chr11 chr12 chr13 chr14 chr15 chr16 chr17 chr18 chr19 chr20
chr21 chr22 chrX chrY

#Check CNVpytor correctly detected the genome reference (hg19 or hg38)
./cnvpytor -root myfile.pytor -ls
```

Calculate the histograms of read depth using bins of 10 kbp, perform GC correction and statistics, and partition the data.

```
# Calculate histograms of read depth, GC correction
./cnvpytor -root myfile.pytor -his 10000
# Partition data
./cnvpytor -root myfile.pytor -partition 10000
```

Finally, call the CNV regions and store the results in a tab-separated file that indicates the CNV type, region (chr:start and end), size, and read depth (normalized to 1), followed by statistics and quality indicators.

```
# Call CNV regions
./cnvpytor -root myfile.pytor -call 10000 > calls.10000.tsv
```

The called CNVs can be filtered to eliminate CNVs with little change in depth and can be annotated with VEP, using the same procedures shown in Chapter 8.

CNVpytor can also use as input the VCF file we previously generated with Strelka and calculate the imbalance in alleles (B-allele frequency or BAF). Heterozygous SNVs are visible in half of the reads, so BAF is usually equal to 0.5. If one chromosomal region is duplicated, the heterozygous SNVs will be visible in one-third or two-thirds of the reads (depending if the SNV is on the duplicated region or not), so BAF will be equal to one-third or two-thirds. This provides an additional signal, in addition to the depth itself, for deciding whether a region is a CNV or not.

Let's import the SNVs from the VCF file generated by Strelka. By default, all chromosomes are selected, unless chromosomes are specified by "-chrom" using the same chromosome names as in the VCF file. We can then apply masks and calculate the BAF histograms. It is possible to make a joint call for CNVs combining the information of read depth and BAF. Alternatively, others prefer to call CNVs from read depth only and use the BAF outputs for manual validation.

```
# Import SNV data previously calculated by Strelka
./cnvpytor -root myfile.pytor -snp somatic.snvs.vcf.gz

#Apply masks from 1000 genomes
./cnvpytor -root myfile.pytor -mask_snps

# Generate BAF histograms
./cnvpytor -root file.pytor -baf 10000

# Call CNVs from combined read depths and BAF (prototype)
./cnvpytor -root myfile.pytor -call combined 10000 > calls.combined.10000.tsv
```

The best way to visualize CNVpytor results is to enter the interactive mode by using the -view option.

```
# Enter CNVpytor interactive mode
./cnvpytor -root myfile.pytor -view 10000

# Plot chromosome 1 from bases 1M to 50M
cnvpytor> 1:1M-50M
```

We can visualize the read depth along the tumor genome and the calls (in the squared dark line) made by CNVpytor. Figure 12.4 shows a deletion and a duplication. For a detailed CNV study using the graphs, we find it more convenient to download the myfile.pytor file, and we can explore it using a locally installed version of CNVpytor.

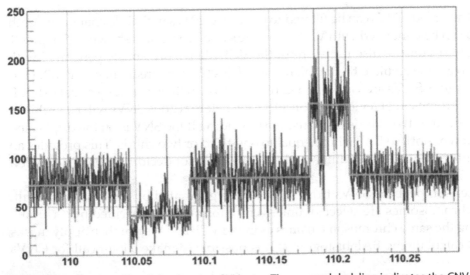

Figure 12.4: Plotting of read depth using CNVpytor. The squared dark line indicates the CNV calls made by CNVpytor.

Measuring Tumor Genomic Instability

We saw that some cancers are caused by mutations in DNA repair genes, whether germline mutations (leading to familial cancer predisposition) or, most commonly, somatic mutations (progressively developed in the tumor). It is important to identify the presence of DNA repair defects as it can change the type of cancer treatment proposed.

There are two ways of identifying DNA repair defects.

- By finding the exact cause of the defect, such as mutation, aberrant epigenetic silencing, or abnormal regulation of DNA repair genes
- By analyzing the overall effect of these mutations on the genome, such as specific patterns of mutations that accumulate in the genome as a result

Mutations or silencing of mismatch repair (MMR) genes, such as the MLH1/2/6 and PMS1/2 family of genes, result in more errors in the most problematic areas of the genome, such as highly repetitive regions of the genome called *microsatellite regions*. These areas are difficult to copy by DNA polymerases during cell replication, and defective MMR genes fail to correct copying errors. This results in *microsatellite instability* (MSI). MSI is most common in colorectal cancer, gastric cancer, and endometrial cancer, and it makes tumors more susceptible to responding to immune checkpoint inhibitors. We will see in this section how to measure the degree of MSI.

Mutations or aberrations in the homologous recombination (HR) pathway, which largely relies on the BRCA1 and BRCA2 genes, causes patterns of loss of heterozygosity, large-scale state transitions, and telomeric allelic imbalances. Patients with cancer and defects in homologous recombination can benefit from PARP inhibitors. The HR genome patterns are more complex to identify and will not be covered here. For more information, look at the HRDetect method implemented in the Signature Lib package at github.com/Nik-Zainal-Group/signature.tools.lib.

Microsatellite instability can easily be identified from BAM files. We will use the popular Microsatellite Analysis for Normal-Tumor inStability (MANTIS) package, which can be downloaded from github.com/OSU-SRLab/MANTIS.

First, use the tool RepeatFinder, which is included in the package, to identify the microsatellites of the genome reference and save their coordinates in a BED file.

```
./RepeatFinder -i $ref_fasta -o msi_loci.bed
```

Second, run MANTIS to determine the instability score between the germline BAM (normal.bam) and the somatic BAM (tumor.bam), using the MSI coordinates calculated in the previous step. Specify the number of threads, for parallelized execution, with the option --threads.

```
python mantis.py \
--bedfile  msi_loci.bed \ # MSI coordinates calculated in previous step
--genome $ref_fasta     \ # genome reference used for the BAM files
-n normal.bam           \ # germline BAM
-t tumor.bam            \ # somatic BAM
--threads 8             \ # nb of threads depending on type of AWS instance
-o myfile.txt             # output file for MSI score.
```

MANTIS produces several output files detailing the microsatellite locus-specific calculations used by the algorithm. The important output file is the `myfile` `.status` file, which contains the final MANTIS result and MSI score. The higher the score is, the more likely the tumor is to have high MSI status. The `myfile` `.status` file also contains the cutoff values more commonly used by clinicians to call a high MSI status. The FDA approved in 2017 the usage of the immune checkpoint inhibitor pembrolizumab for all unresectable or metastatic cancers with high MSI or MMR deficiency, irrespective of the tissue of origin.

Summary

We have arrived at the end of our final chapter. We have covered a lot of ground together, including exploring molecular biology, sourcing a genome, discovering how to best set up our environment on the cloud, analyzing genomes, and expanding to other applications and even proteins. It is an endless journey, where the more we know, the more we discover what we do not know. Genomes are incredibly complex as they are like legacy machines that have accumulated layers and layers of evolution since the beginning of life on Earth. They are incredibly redundant, but it is this redundancy that makes them robust. After analyzing so many genomes, it feels incredible to see so many mutations in completely healthy individuals, while at the same time, and in very rare cases, a misspelled single letter can bring disarray. Genomes can give us insights about our predispositions but, with rare exceptions, can enable us to take action to avoid developing these conditions. Our genomes also make us humble as we share so many genes with very simple life forms and open a window in the process of evolution. The function of half of our genes is still unknown, and you might be the one who will make new discoveries. There is also an enormous gap between what we know of the genome and how to apply this knowledge to benefit patients. In addition to sequencing your own family genomes, there are a bewildering number of public databases and published data that can be analyzed, and the power of the cloud gives the computing power to everyone to dive into it. Do not be intimidated by the scientific literature. We still know very little, and most things are yet to be discovered. Above all, and in all your genomic adventures in the cloud, have fun!

Notes

Hanahan, D., & Weinberg, R. A. The hallmarks of cancer. Cell. 2000;100(1):57-70. doi:10.1016/s00928z674(00)81683-9

Hanahan, D., & Weinberg, R. A. Hallmarks of cancer: the next generation. Cell. 2011;144(5):646-674. doi:10.1016/j.cell.2011.02.013

Hanahan, D. Hallmarks of cancer: new dimensions. Cancer Discov. 2022; 12(1):31-46. doi:10.1158/2159-8290.CD-21-1059

Jonsson, H., et al. Differences between germline genomes of monozygotic twins. Nat Genet. 2021;53(1):27-34. doi:10.1038/s41588-020-00755-1

Martincorena, I., et al. Tumor evolution. High burden and pervasive positive selection of somatic mutations in normal human skin. Science. 2015;348(6237):880-886. doi:10.1126/science.aaa6806

Notes

Hanahan, D. & Weinberg, R. A. The hallmarks of cancer. Cell 100, 57–70 doi:10.1016/s0092-8674(00)81683-9

Hanahan, D. & Weinberg, R. A. Hallmarks of cancer: the next generation. Cell 144, 646–674, doi:10.1016/j.cell.2011.02.013

Hanahan, D. Hallmarks of cancer: new dimensions. Cancer Discov 12(1), 31–46, doi:10.1158/2159-8290.CD-21-1059

Loeb, L. A. et al. A mutator phenotype in cancer. Cancer Res 61, 3230–3239 DOI: not available. Issue date 1999-04-15

Tomasetti, C. et al. Only three driver gene mutations are required for the development of lung and colorectal cancers. Proc Natl Acad Sci U S A 112, 118–123, doi:10.1073/pnas.1421839112

Index